AI Interviews with
ASTRONAUT
LEGENDS
Voices from the Space Age Beyond Time
CCCP
Dr. Borko Furht

AI Interviews with Astronaut Legends

Voices from the Space Age Beyond Time

Table of Contents

Part II – The Journey to the Moon

Part III – Courage, Discovery, and Leadership

Part IV – International Explorers of Space

Preface

Why Humanity Reaches for the Stars

From the earliest days of civilization, humans have looked toward the night sky with curiosity, wonder, and ambition. The stars have always represented both mystery and possibility. Ancient astronomers mapped their patterns, philosophers imagined worlds beyond Earth, and poets dreamed of journeys among the constellations. Yet for most of human history, the idea of leaving our planet remained only a dream.

The twentieth century transformed that dream into reality. In the span of only a few decades, humanity progressed from fragile experimental rockets to orbiting the Earth, walking on the Moon, and constructing laboratories in space. These achievements were not simply technological milestones, they were profound expressions of human courage, curiosity, and determination.

Behind every mission stood individuals willing to confront enormous risks in pursuit of discovery. Astronauts and cosmonauts climbed aboard machines that had never been fully tested, ventured into environments no human had experienced, and trusted both science and their own bravery to carry them safely beyond Earth. Some returned as heroes.

Some paid the ultimate price. Their stories form one of the most remarkable chapters in the history of human exploration.

This book presents imagined conversations with many of those pioneers—men and women whose journeys expanded the boundaries of human experience. Through the assistance of artificial intelligence, we recreate interviews that explore their motivations, struggles, triumphs, and reflections on humanity's quest to reach the stars.

These dialogues are not literal historical transcripts. Rather, they are thoughtful reconstructions grounded in historical records, biographies, and the documented words of these astronauts. Artificial intelligence allows us to synthesize this knowledge and imagine how these remarkable individuals might respond if they could speak to us today.

The goal of this book is not only to remember their achievements but also to reflect on the deeper meaning of exploration. Why do humans risk so much to travel into the unknown? What drives individuals to climb into rockets and leave their home planet behind? And what lessons can we learn from their courage?

Space exploration represents one of humanity's greatest collective endeavors. It unites science, engineering, imagination, and the timeless human desire to discover what lies beyond the horizon.

The astronauts featured in these pages remind us that progress is never easy. It requires vision, sacrifice, and an unwavering belief that the unknown is worth exploring. Their journeys did not end with the missions they flew. Their legacy continues to inspire scientists, engineers, dreamers, and explorers around the world. As humanity prepares for new missions to the Moon, Mars, and beyond, their voices remain as relevant as ever. They remind us that exploration is not merely about reaching distant destinations, it is about expanding the limits of what humanity believes is possible.

Introduction

AI Conversations with the Pioneers of Space Exploration

The story of space exploration is one of the most extraordinary adventures in human history. Only a little more than sixty years ago, no human had ever left Earth. The idea of traveling into space existed primarily in science fiction novels and visionary sketches by early rocket scientists.

Then, in 1961, a young Soviet pilot named **Yuri Gagarin** became the first human to orbit the Earth. His flight lasted just 108 minutes, yet it changed the course of history. For the first time, a human being had seen our planet from space—a fragile blue sphere suspended in the vast darkness of the cosmos.

Soon afterward, American astronauts followed. **Alan Shepard** made the first American journey into space, and **John Glenn** orbited the Earth, becoming a national hero. These early missions marked the beginning of a fierce yet inspiring competition between nations, the Space Race.

But the Space Race was about more than geopolitics. It was about pushing the limits of science and engineering. It was about proving that humans could survive—and thrive—beyond the atmosphere of our home planet. The achievements came rapidly.

Astronaut **Alexei Leonov** performed the first spacewalk, floating freely outside his spacecraft. Just a few years later, **Neil Armstrong** stepped onto the surface of the Moon, uttering the famous words that would echo through history: *"That's one small step for man, one giant leap for mankind."*

These triumphs required extraordinary courage. Early spacecrafts were fragile, experimental, and often unpredictable. Astronauts trusted not only technology but also their own skill and composure in situations where failure could mean disaster.

And sometimes, tragedy occurs. The history of space exploration includes moments of profound loss—from the **Apollo 1** fire that claimed the lives of astronauts, including **Roger B. Chaffee**, to the **Challenger** and **Columbia** disasters that shocked the world. These tragedies remind us that exploration always carries risk.

Yet even in the face of danger, humanity continued to reach for the stars. The astronauts featured in this book represent many different nations, eras, and missions. Some were pioneers of the earliest spaceflights. Others walked on the Moon. Some helped build the foundations of modern space science through the Space Shuttle program and international cooperation. Together, their stories form a powerful narrative of human determination.

In this book, artificial intelligence serves as a creative tool that allows us to revisit their lives and achievements through imagined interviews. Drawing from historical records, speeches, memoirs, and documented experiences, these conversations attempt to capture the spirit, perspective, and personality of the astronauts who helped shape the Space Age.

Through these dialogues, we explore questions that remain timeless:

- What does Earth look like from space?
- What does it feel like to leave the planet behind?
- How do astronauts confront the possibility of danger or death?
- How does seeing our world from orbit change the way one thinks about humanity?

Perhaps most importantly, these conversations remind us that space exploration is not only about rockets and technology. It is about people. People who dared to imagine what had never been done. People who accepted immense risk in the pursuit of discovery. People who believed that humanity's future lies not only on Earth but also among the stars.

Their journeys inspire us to ask a simple yet profound question: Where will humanity go next? The answer may lead us back to the Moon, onward to Mars, and eventually far beyond our solar system.

But whatever path the future takes, it will always build upon the courage and vision of the pioneers whose voices we revisit in this book. Their legacy continues to guide humanity's journey into space. And their stories remind us that the greatest frontier of all still lies ahead.

Epigraph

"That's one small step for man, one giant leap for mankind."

— Neil Armstrong

"We choose to go to the Moon in this decade and do the other things, not because they are easy, but because they are hard."

— John F. Kennedy

"For the first time, I saw the horizon as a curve. The beauty of Earth was overwhelming."

— Yuri Gagarin

"Somewhere, something incredible is waiting to be known."

— Carl Sagan

"The Earth is the cradle of humanity, but mankind cannot stay in the cradle forever."

— Konstantin Tsiolkovsky

Yuri Gagarin

Alan Shepard

John Glenn

Gus Grissom

Alexei Leonov **Neil Armstrong**

Michael Collins **Pete Conrad**

Edgar Mitchell **Gene Cernan**

Jim Lovell **Roger Chaffee**

Sally Ride **Judith Resnik**

Vladimir Komarov **Georgy Dobrovolsky**

Victor Patsayev **Vladislav Volkov**

Christa McAuliffe **Ronald McNair**

Kalpana Chawla

Rick Husband

Interview with Yuri Gagarin

The First Human Journey into Space

Introduction

On the morning of April 12, 1961, humanity crossed a boundary that had existed since the beginning of our species. For thousands of years, humans had looked up at the night sky and wondered what lay beyond the blue ceiling of Earth's atmosphere. Astronomers studied the stars, poets wrote about distant worlds, and dreamers imagined traveling among the planets. Yet no human being had ever left the surface of Earth.

That changed when a young Soviet pilot named Yuri Gagarin climbed into a small spherical spacecraft called ***Vostok 1****. The spacecraft was barely large enough for a single person. Surrounded by instruments and strapped tightly in his seat, Gagarin prepared to become the first human in history to leave the planet. Outside the capsule stood a towering rocket designed to carry him into an environment that had never been experienced by a human being.*

No one knew exactly what would happen. Scientists were unsure how the human body would react to weightlessness.

Engineers could only estimate the stresses the spacecraft might experience during launch and reentry. Every moment of the mission carried uncertainty.

Yet the launch proceeded. At 9:07 a.m. local time, the engines of the rocket ignited with thunderous force. As the vehicle rose into the sky above the Baikonur Cosmodrome, Gagarin spoke a single word that would become legendary in the history of space exploration:

"Poyekhali!" — "Let's go!" Within minutes, he had left Earth's atmosphere and entered orbit. For the next 108 minutes, Yuri Gagarin circled the planet, becoming the first human to see Earth from space. Through the small window of his spacecraft, he watched continents, oceans, and clouds drift silently beneath him. The curvature of the Earth was visible against the darkness of space, revealing the fragile beauty of the world humanity calls home.

His flight was short, but its impact was enormous. Gagarin's journey marked the true beginning of human space exploration. It demonstrated that humans could travel beyond Earth and survive in the harsh environment of space. It opened the door to future missions that would eventually lead astronauts to walk on the Moon and explore the far reaches of the solar system.

Overnight, Yuri Gagarin became an international symbol of courage, curiosity, and human achievement. His smile and humble personality made him one of the most admired figures of the Space Age. But more importantly, his historic flight proved that the dream of leaving Earth was no longer just imagination, it had become reality. Yuri Gagarin's journey lasted less than two hours. Yet in those 108 minutes, humanity's future changed forever.

In the following conversation, we revisit the reflections of the man who first carried humanity beyond the boundaries of our planet and opened the path to the stars. 🚀

Yuri Gagarin portrait Source: Wikimedia Commons - CC BY 4.0

Yuri Alekseyevich Gagarin (9 March 1934 – 27 March 1968) was a Soviet pilot and cosmonaut who, aboard the first successful crewed spaceflight, became the first person to journey into outer space. Travelling on Vostok 1, Gagarin completed one orbit of Earth on 12 April 1961, with his flight taking 108 minutes. By achieving this major milestone for the Soviet Union amidst the Space Race, he became an international celebrity and was awarded many medals and titles, including his country's highest distinction: Hero of the Soviet Union.

---

1. Yuri Alekseyevich, when did you first become fascinated with aviation and the idea of flying?

My fascination with flight began when I was a young boy. During the difficult years after the war, aviation represented hope and progress for our country. I remember watching aircraft flying overhead and imagining what it would feel like to sit in the cockpit. Later, when I began training as a pilot, that dream became real. At the time, none of us imagined that we might one day fly beyond the atmosphere of Earth.

---

2. What was your reaction when you were selected to become the first human to fly into space?

It was an extraordinary honor, but it was also a tremendous responsibility. The Soviet space program had many talented pilots and engineers working together. When I learned that I would fly aboard Vostok 1, I felt both excitement and determination. I knew that the mission was not only about me—it represented the efforts of thousands of people who had made this achievement possible.

Gagarin in his Vostok 1 spacesuit on 12 April 1961
Source Wikimedia Commons - CC BY 4.0

---

3. What do you remember about the moment of launch on April 12, 1961?

The launch was powerful and unforgettable. When the rocket engines ignited, the entire spacecraft vibrated with tremendous force. As the rocket climbed higher, I felt the weight of gravity pressing me into my seat. Yet I also felt incredible excitement, knowing that humanity was leaving Earth for the first time.

---

4. When you first saw Earth from space, what did you feel?

The view was breathtaking. I saw our planet as a beautiful blue sphere surrounded by darkness. The horizon appeared curved, and the atmosphere looked thin and delicate. At that moment I realized how small and precious our world truly is. It was one of the most profound experiences of my life.

Gagarin Monument - Poznan.
Source: Wikimedia Commons – CC BY-SA 4.0

---

5. Did you feel fear during your historic flight?

Every new mission involves uncertainty. We were exploring an environment that no human had experienced before. However, the training we received and the confidence we had in our engineers helped us focus on the mission rather than the risks. I believed deeply in the program and in the people who made the spacecraft possible.

---

6. Your flight lasted only 108 minutes, yet it changed history. Did you understand the significance at the time?

During the flight, my primary task was to observe and report on the spacecraft's performance and my physical condition. Only after returning safely to Earth did, I begin to understand the enormous impact the mission had on the world. Humanity had taken its first step beyond our planet.

---

7. How did your life change after becoming the first human in space?

Life changed dramatically. I traveled to many countries and met people from all over the world. Everywhere I went, I sensed that people felt inspired by the idea of space exploration. It showed me that spaceflight could unite humanity through a shared sense of wonder.

---

8. Colonel Gagarin, after your historic flight aboard Vostok 1, you became an international hero and were invited to visit more than thirty countries. During those travels you met many world leaders and millions of people around the world. Which countries and leaders left the strongest impression on you, and what do you remember most from those visits?

The period following my flight was truly extraordinary. I suddenly found myself traveling across the world, visiting many countries and meeting people from different cultures who shared a common fascination with space exploration.

What impressed me most was not only the leaders I met, but the enthusiasm and warmth of ordinary people everywhere. Whether in Europe, Asia, or the Americas, crowds gathered to celebrate what they saw as a milestone for all humanity, not just for one nation.

In the United Kingdom, I had the honor of meeting **Queen Elizabeth II**. She welcomed me with great warmth and curiosity about the experience of spaceflight. I was struck by how interested she was in the scientific and technical aspects of the mission.

During my visits to several countries in Asia and Africa, I was deeply moved by the excitement among young people and students. Many told me that seeing a human travel into space made them believe that science and education could open new possibilities for their own futures.

I also met leaders of many nations who viewed the achievement of spaceflight as an important step for human progress. Even during the tense years of the Cold War, it was clear that the exploration of space had the power to inspire cooperation and admiration across political boundaries.

One of the most memorable aspects of those journeys was realizing that my flight was seen not only as a Soviet accomplishment but as a milestone for all humankind. People everywhere looked toward the sky and imagined that someday they too might reach the stars.

That was perhaps the most rewarding part of the experience: seeing how a single flight could unite people around the world in a shared sense of wonder about the universe.

Ten ruble coin, Bank of Russia - 40 years of space flight Y.A. Gagarin

---

9. Colonel Gagarin, a famous quote often attributed to you says that during your spaceflight you remarked, *"I don't see any God up here."* This statement has been widely repeated over the years. Could you clarify whether you said this during your mission and share your thoughts about it?

That phrase has often been attributed to me, but I did **not** make that statement during my flight aboard **Vostok 1**.

During the mission I was focused entirely on operating the spacecraft, communicating with ground control, and observing the extraordinary view of Earth from orbit. The flight lasted only 108 minutes, and every moment required careful attention to the mission tasks.

The quote that is often repeated actually originated later in political speeches during the Cold War era. At that time, space exploration was closely connected with ideological rivalry, and statements about science and religion were sometimes used for political messaging. Over time, the phrase became incorrectly associated with my flight.

What I truly remember from my journey into space was the beauty of our planet. From orbit, Earth appeared as a magnificent sphere surrounded by the darkness of space. The oceans, clouds, and continents formed a breathtaking view that left a deep impression on me.

Many astronauts and cosmonauts have described a similar feeling when seeing Earth from space—a sense of wonder and reflection about humanity's place in the universe. Spaceflight does not answer philosophical or spiritual questions about life. Instead, it reminds us how small and precious our world is and how important it is for humanity to live together peacefully on this single planet. That, perhaps, is the most meaningful lesson that space exploration can offer. 🚀

---

10. What message would you share about international cooperation in space?

Space belongs to all humanity. Although the early years of space exploration were shaped by competition between nations, I believe that the exploration of the cosmos should ultimately bring people together. The universe is vast enough for all nations to explore together.

Statue of Yuri Gagarin in Tagus Park
Source: Wikimedia Commons – CC0

11. Colonel Gagarin, your flight aboard Vostok 1 opened the era of human space exploration. Looking ahead, what do you hope humanity will achieve in space in the future?

My flight was only the very first step in a much larger journey for humanity. When we launched into space in the early 1960s, we were just beginning to learn how humans could travel beyond Earth.

I believe the future of space exploration will be far more ambitious than anything we imagined at the beginning of the Space Age.

One of the most important goals will be continued exploration of the Moon. The Moon can serve as a scientific laboratory and perhaps even as a steppingstone for deeper exploration into the solar system.

Beyond the Moon, I believe humanity will eventually send astronauts to **Mars**. Such a mission will require extraordinary cooperation between scientists, engineers, and nations around the world, but it would represent one of the greatest achievements in human history.

Another important development will be the construction of long-term space stations where astronauts from different countries can live and work together while conducting scientific research. These stations can help us better understand how humans can survive and work in space for extended periods.

Perhaps the most inspiring aspect of future exploration will be international cooperation. When people look at Earth from space, they do not see borders or political divisions; they see one beautiful planet shared by all humanity. For that reason, I hope that future missions will increasingly involve cooperation between many nations. Exploration of the cosmos should be a project for all mankind.

The universe is vast, and humanity has only begun to explore it. If we continue to pursue knowledge, science, and peaceful cooperation, I believe that future generations will travel farther than we ever imagined—and perhaps one day even reach the stars. 🚀

Gagarin with U.S. Vice President Hubert Humphrey, French Prime Minister Georges Pompidou and Gemini 4 astronauts Jim McDivitt and Ed White at the 1965 Paris Air Show

---

12. Colonel Gagarin, as the first human to journey into space aboard Vostok 1, your flight opened a new chapter in human history. If you could speak directly to future generations who look toward the stars and dream of exploring space, what message would you share with them?

To the future explorers of space, I would say this: never stop being curious about the universe. Human progress has always been driven by the desire to explore the unknown. From the first sailors who crossed oceans to the scientists who built rockets capable of leaving Earth, each generation has taken another step toward understanding our place in the cosmos.

When I looked at Earth from space during my flight, I saw a beautiful and fragile planet surrounded by the vast darkness of the universe. From that perspective, our world appeared small, yet incredibly precious. That view reminds us that humanity shares one home and one destiny.

The explorers of the future will travel farther than we did in the early days of spaceflight. You will build stations around the Moon, send missions to Mars, and perhaps one day journey even farther into the solar system. But as you explore new worlds, remember that exploration is not only about technology or achievement. It is also about cooperation, curiosity, and the desire to expand human knowledge.

Work together across nations and cultures. Protect the planet that gave you life. And continue to look toward the stars with courage and imagination. The universe is vast, and humanity has only just begun its journey. 🚀

Yuri Gagarin statue at the Royal Greenwich Observatory in London, England

Interview with Alan Shepard

First American in Space

Introduction

On May 5, 1961, just twenty-three days after the Soviet Union sent Yuri Gagarin into orbit, the United States prepared to answer with a bold step of its own. At Cape Canaveral in Florida, a slender rocket stood on the launch pad, carrying a small capsule named ***Freedom*** *7. Inside sat a naval test pilot known for his calm confidence and sharp technical skill: Alan Shepard.*

Unlike the powerful rockets that would later carry astronauts to the Moon, the Mercury-Redstone rocket was modest in size. The spacecraft itself was barely large enough for one person. Yet the mission represented something far greater than a brief journey into space. It was America's first attempt to send a human beyond Earth's atmosphere.

Shepard had trained intensely for this moment. Like the other members of NASA's original Mercury Seven astronaut corps, he was a test pilot accustomed to pushing experimental aircraft to their limits. Courage, precision, and discipline defined his profession. When the countdown reached zero, the rocket's engines ignited with a thunderous roar, lifting Freedom 7 into the sky. Within minutes, Shepard became the first American to travel into space.

His flight lasted only fifteen minutes, but it changed the course of American space exploration. Shepard demonstrated that astronauts could control their spacecraft, endure the stresses of launch and reentry, and safely return to Earth.

Years later, Shepard would once again make history—this time as the commander of ***Apollo 14****, becoming the fifth human to walk on the Moon. From the earliest days of the Mercury program to the triumphs of Apollo, Alan Shepard helped lead America's journey into space.*

In the following conversation, we revisit the reflections of the man who opened the path for American astronauts to reach the stars.

Alan Bartlett Shepard Jr. (November 18, 1923 – July 21, 1998) was an American astronaut. In 1961, he became the second person and the first American to travel into space and, in 1971, he became the fifth and oldest person to walk on the Moon, at age 47.

---

1. Commander Shepard, when did your interest in aviation begin?

My interest in aviation started during my youth when airplanes were still relatively new and exciting machines. By the time I entered the U.S. Naval Academy, I already knew that I wanted to become a pilot. Flying was both challenging and rewarding, and test piloting later allowed me to push the boundaries of aviation technology.

Shepard's 105th F4U sortie on USS Franklin D. Roosevelt

---

2. What was your reaction when NASA selected you as one of the original Mercury astronauts?

It was an incredible opportunity. The Mercury program represented the first step in America's effort to explore space. All seven of us understood that the missions would be difficult and that much was still unknown. But we also knew we were part of something historic.

---

3. What do you remember about the moments just before the launch of Freedom 7?

The launch sequence involved many hours of preparation and waiting. By the time the countdown reached zero, I was more focused on the procedures than the excitement of the moment. However, when the rocket finally lifted off, there was no doubt that something extraordinary was happening.

The Mercury Seven astronauts with a USAF F-106. From left to right: Scott Carpenter, Gordon Cooper, John Glenn, Gus Grissom, Wally Schirra, Alan B. Shepard and Deke Slayton.

4. Your flight lasted only about fifteen minutes. Did it still feel like a significant milestone?

Absolutely. Even though the mission was suborbital and relatively short, it demonstrated that humans could travel into space and operate a spacecraft. That mission helped build confidence in technology and paved the way for longer and more ambitious missions.

---

5. Commander Shepard, Yuri Gagarin completed a full orbital flight aboard the Vostok spacecraft that lasted about 108 minutes, while your historic flight aboard Freedom 7 lasted only about 15 minutes and followed a suborbital trajectory. It is also often noted that Gagarin's spacecraft operated almost entirely in automatic mode, whereas you had some manual control over your capsule. Could you elaborate on these differences between the two missions?

That's a very good question, and it reflects the different stages of development in the early space programs of the Soviet Union and the United States.

Gagarin's flight aboard **Vostok 1** was indeed a full **orbital mission**. His spacecraft reached orbital velocity and circled the Earth once before reentering the atmosphere. My mission aboard **Freedom 7**, part of the Mercury program, was a

suborbital flight, which means the spacecraft traveled into space and then followed a ballistic trajectory back to Earth without completing an orbit.

The difference was largely due to the rockets that were available at the time. The Soviet Union had developed a powerful launch vehicle capable of placing the Vostok spacecraft into orbit. In the United States, our first crewed missions used the **Redstone rocket**, which was not powerful enough to achieve orbital velocity. As a result, the Mercury program began with suborbital flights before progressing to orbital missions later that year.

Another interesting difference involved the **control systems**. The Vostok spacecraft was designed to operate almost entirely in **automatic mode**. Soviet engineers were concerned about how a human might react to weightlessness and therefore minimized the amount of manual control available to the pilot during the mission.

In contrast, the Mercury spacecraft gave the astronaut **some ability to control the capsule**. During my flight I could monitor spacecraft systems, adjust orientation using the reaction control thrusters, and interact with the spacecraft's controls if necessary.

In practice, however, both missions relied heavily on automated systems and guidance from ground control. The spacecraft were still relatively simple compared with modern vehicles, and their primary objective was to demonstrate that humans could travel safely into space and return.

Both flights were historic in their own ways. Gagarin's mission proved that humans could survive an orbital journey around Earth, while the Mercury flights demonstrated that astronauts could actively participate in spacecraft operations.

Together, these early missions helped lay the foundation for the much more complex spaceflights that followed, eventually leading to the Apollo missions and the exploration of the Moon. 🚀

---

6. Admiral Shepard, when you became the first American to travel into space aboard Freedom 7 in 1961, the experience was completely new for any U.S. astronaut. During that historic journey, what surprised you the most?

One of the things that surprised me most was how smooth and controlled the flight felt once we reached space. During launch, of course, there was tremendous acceleration as the rocket lifted off. The Redstone rocket produced a powerful push that you could clearly feel in your seat. But once the spacecraft reached its trajectory and the engines shut down, the environment became remarkably quiet and calm.

Experiencing **weightlessness** for the first time was also fascinating. Everything inside the capsule seemed to float effortlessly, and it gave you a completely different sense of movement compared to anything we experience on Earth.

Another unforgettable moment was looking out of the spacecraft window and seeing **the curvature of the Earth**. The view of the planet against the blackness of space was extraordinary. Even though my flight lasted only about fifteen minutes, that perspective left a lasting impression.

What also surprised me was how quickly everything happened. Years of training and preparation led up to a mission that unfolded in just a few minutes. Yet during those moments we were demonstrating something extremely important—that humans could travel into space, operate a spacecraft, and return safely.

In many ways, the flight confirmed that the technology and the training had worked exactly as intended. That success helped open the door for the longer and more complex missions that followed in the years ahead. 🚀

---

7. The Mercury astronauts were test pilots accustomed to risk. How did that background help prepare you for spaceflight?

Test pilots are trained to deal with uncertainty and complex systems. We were used to evaluating new technology and responding quickly when something unexpected happened. That experience was extremely valuable during the early days of space exploration.

Marine Corps HUS-1 helicopter from HMR-262 retrieves Freedom 7 from the Atlantic.

---

8. Years after your Mercury flight, you walked on the Moon during Apollo 14. How did that experience compare?

Walking on the Moon was an entirely different experience. The Apollo missions represented a tremendous leap forward in technology and capability. Standing on the lunar surface and looking back toward Earth was one of the most extraordinary moments of my life.

---

9. Commander Shepard, during your remarkable career you participated in several major NASA flight programs, including Mercury, which sent the first Americans into space, and Apollo, which ultimately carried astronauts to

the Moon. Of all the NASA programs in which you were involved, which did you find the most exciting, and why?

Each of the programs I was involved in had its own excitement and importance, because they represented different stages in the development of human spaceflight.

My first mission was part of **Project Mercury**, where I piloted the spacecraft **Freedom 7** in 1961. That flight made me the **first American to travel into space**, and its primary goal was to demonstrate that a human could safely travel into space and return to Earth. At that time, everything we were doing was completely new. We were testing the limits of both technology and human capability in the space environment.

However, if I had to choose the program that was most exciting and meaningful to me personally, I would say the **Apollo program**, particularly **Apollo 14**. Apollo represented the culmination of everything NASA had been working toward since the early days of Mercury. Instead of simply proving that humans could survive in space, we were now traveling nearly **400,000 kilometers to the Moon**, landing on its surface, conducting scientific exploration, and returning safely to Earth. During Apollo 14 in 1971, my crewmate Edgar Mitchell and I spent more than **33 hours on the lunar surface**, performing experiments and collecting geological samples.

Standing on the Moon and looking back at Earth is something difficult to describe. At that moment you realize that all the years of work—from the earliest Mercury flights to the enormous engineering effort behind Apollo—had led to that single achievement.

Mercury was the beginning of America's journey into space, and it carried tremendous excitement because we were pioneers. But Apollo was the program that truly expanded the horizon of human exploration. So, while Mercury will always hold a special place in history—and in my heart—Apollo was, without question, the most extraordinary experience of my career. It allowed humanity not just to reach space, but to set foot in another world.

Shepard and his wife Louise meet First Lady Jacqueline Kennedy, President John F. Kennedy and Vice President Lyndon B. Johnson at the South Portico of the White House, prior to his receiving the NASA Distinguished Service Medal on May 8, 1961.

---

10. During Apollo 14, you famously hit a golf ball on the Moon. What inspired that moment?

Well, astronauts occasionally appreciate a bit of humor. I brought along a small golf club head and attached it to a sampling tool. It was a lighthearted moment that showed even astronauts could enjoy a little fun while exploring another world.

---

11. Admiral Shepard, you were one of the original astronauts of the Mercury program and later commanded Apollo 14, becoming the fifth human to walk

on the Moon. Looking back at the early years of space exploration, what do you believe was the greatest achievement of those pioneering space programs?

The greatest achievement of the early space programs was proving that **human beings could travel safely beyond Earth and explore space**.

When Project **Mercury** began, spaceflight was still largely theoretical. Scientists and engineers did not yet know how the human body would respond to weightlessness, how spacecraft systems would function in orbit, or whether astronauts could effectively control vehicles outside Earth's atmosphere.

The Mercury missions demonstrated that humans could launch into space, survive the extreme conditions of flight, operate spacecraft systems, and return safely to Earth. That may seem straightforward today, but at the time it was a major step forward in science and engineering.

Those early flights laid the foundation for everything that followed. The knowledge gained from Mercury allowed NASA to develop the more advanced **Project Gemini**, which introduced techniques such as rendezvous and docking in orbit. Those capabilities were essential for the success of the **Apollo program**, which ultimately carried astronauts to the Moon.

Another important achievement was the way these programs inspired people around the world. Millions of young students became interested in science, engineering, and technology after watching the early space missions.

In short, the early space programs did more than send astronauts into space—they opened a new frontier for humanity and demonstrated what can be accomplished through determination, scientific innovation, and teamwork.

The crew of Apollo 14: Edgar Mitchell, Shepard and Stuart Roosa

---

12. Commander Shepard, your pioneering role in the early days of human spaceflight made you one of the most recognizable figures of the Space Age. Over the years, your legacy has even appeared in popular culture. For example, archival footage of you is included in the opening montage of the television series *Star Trek: Enterprise*, which highlights the history of human exploration. In addition, some science-fiction works have created characters believed to be inspired by you, such as

Alan Tracy in *Thunderbirds* and Commander Shepard in the video game series *Mass Effect*. Could you share your thoughts on how space pioneers have influenced science fiction and popular culture?

It's always interesting to see how real achievements in exploration find their way into popular culture. Human spaceflight captured the imagination of people around the world during the early 1960s, and it naturally influenced writers, filmmakers, and game designers who were thinking about the future.

When I made the **Freedom 7** flight in 1961, space travel was still something very new and exciting. For many people, the idea of humans traveling beyond Earth had previously existed only in science-fiction stories. Once astronauts actually began flying into space, it gave those stories a new sense of realism.

The inclusion of historical footage in programs like *Star Trek: Enterprise* is a good example of how creators like to connect fictional exploration with real milestones in human history. It reminds viewers that the journey toward exploring other worlds did not begin with starships—it began with small rockets, early spacecraft, and the efforts of many dedicated engineers and astronauts.

As for fictional characters inspired by early astronauts, I think that reflects how strongly the Space Age influenced popular imagination. Shows like *Thunderbirds* and later works such as the *Mass Effect* games often portray explorers who are courageous, technically skilled, and willing to take risks to expand humanity's reach into space.

In reality, of course, space exploration is never the work of a single hero. Every astronaut represents thousands of people who design the spacecraft, build the rockets, analyze the data, and support the missions from the ground.

If our early flights helped inspire writers, artists, and storytellers to imagine humanity exploring the stars, then I think that's a wonderful outcome. Exploration has always inspired creativity, and science fiction often encourages future scientists and engineers to pursue those dreams.

In many ways, imagination and exploration go hand in hand—the stories we tell about the future often help inspire the discoveries that make that future possible. 🚀

---

13. Admiral Shepard, as one of the pioneers of human spaceflight and a commander who walked on the Moon during Apollo 14 mission, what message would you share with the future explorers who will one day travel to the Moon again and perhaps even to Mars?

To the explorers of the future, I would say that you are continuing one of the greatest adventures in human history. When we flew the early missions of the **Project Mercury**, we were taking the very first steps into space. Later, during the **Apollo program**, we demonstrated that humans could travel nearly a quarter of a million miles from Earth and land on another world.

Those missions were only the beginning. The explorers who return to the Moon and travel onward to Mars will face challenges that are even greater than those we encountered. You will travel farther from Earth, remain in space for longer periods, and rely on new technologies that we could only imagine.

But the spirit of exploration remains the same. Prepare thoroughly, trust your training, and rely on the teamwork of the thousands of scientists, engineers, and technicians who make these missions possible. No astronaut ever reaches space alone—every flight represents the efforts of an entire community dedicated to discovery.

I would also encourage future explorers to remember the perspective that spaceflight provides. When you look back at Earth from the Moon or from deep space, you see a small and beautiful planet that all humanity shares. Protect that world and work together across nations to continue exploring the universe.

If future generations carry forward that spirit of curiosity, courage, and cooperation, then humanity's journey beyond Earth will continue to grow—and the horizons of exploration will expand far beyond anything we accomplished in the early days of the Space Age.

Shepard poses next to the American flag on the Moon during Apollo 14. Source/Author: NASA – Public Domain

Interview with John Glenn

First American to Orbit Earth

Introduction

In the early 1960s, the world was gripped by a new kind of competition—one not fought on land or sea, but in the vast frontier of space. Just months after the Soviet Union sent the first human into orbit, the United States was determined to prove that it too could send astronauts safely around the Earth. The challenge was immense. Rockets were powerful but unpredictable. Spacecraft were experimental. No one knew exactly how the human body would respond to prolonged weightlessness.

Into this uncertainty stepped John Glenn. A decorated Marine Corps pilot and combat veteran, Glenn had already built a reputation as one of America's most skilled aviators. Calm under pressure and deeply committed to his country, he was selected as one of NASA's original ***Mercury Seven astronauts****, the group of pioneers chosen to lead America's first human spaceflight missions.*

On February 20, 1962, Glenn climbed aboard the tiny ***Friendship*** *7 spacecraft perched atop an Atlas rocket at Cape Canaveral. Millions of Americans watched anxiously* *as the countdown reached zero and the rocket thundered into the sky. Within minutes, Glenn became the first American to orbit the Earth.*

During his mission, he circled the planet three times, witnessing breathtaking views of oceans, continents, and glowing city lights far below. His flight lasted nearly five hours and demonstrated that astronauts could operate spacecraft successfully in orbit. But Glenn's story did not end there.

Decades later, at the age of seventy-seven, he returned to space aboard the Space Shuttle ***Discovery****, becoming the oldest person ever to fly in orbit and symbolizing a lifetime dedicated to exploration and service. John Glenn represented a generation of astronauts who helped transform spaceflight from an uncertain experiment into a defining achievement of modern science.*

His courage, discipline, and optimism helped inspire a nation—and proved that the frontier of space was within humanity's reach.

John Herschel Glenn Jr. (July 18, 1921 – December 8, 2016) was an American Marine Corps aviator, astronaut, businessman, and politician. He was the third American in space and the first to orbit the Earth, circling it three times in 1962. Following his retirement from NASA, he served from 1974 to 1999 as a U.S. senator from Ohio. In 1998, he flew into space again at the age of 77.

---

1. Senator Glenn, when did your fascination with flight begin?

My interest in aviation began when I was very young. I remember taking my first airplane ride with my father, and the experience left a lasting impression on me. From that moment forward, I knew that flying was something I wanted to pursue.

---

2. What did it mean to you to be selected as one of NASA's original Mercury astronauts?

Being selected as one of the Mercury Seven was an extraordinary honor. We understood that the program represented the first steps of American human spaceflight. Each of us recognized that the missions would be difficult and that we were venturing into unknown territory.

---

3. What do you remember most about the launch of Friendship 7?

The launch of *Friendship 7* during the Mercury-Atlas 6 mission is something I will never forget. After years of preparation and testing in the NASA Mercury program, that moment on the launch pad felt both historic and intensely personal.

As the countdown reached its final seconds, everything became very quiet inside the capsule. You are strapped into a small spacecraft, lying on your back, surrounded by instruments, and you know that in a few moments a powerful rocket will lift you into space.

When the Atlas rocket engines ignited, the vibration was immediate and intense. You could feel the enormous force building beneath you. Then suddenly the rocket began to rise, and the acceleration pressed you firmly into your seat.

What I remember most clearly is the realization that the mission was truly underway—that we were finally doing something humans had only imagined before: orbiting the Earth. As the rocket climbed higher and the sky grew darker, I felt an overwhelming sense of excitement and responsibility. At that moment, I knew that the mission was not just about one astronaut. It represented the work of thousands of engineers and scientists, and the hopes of a nation determined to explore space.

Glenn standing in the cockpit of a F-106B in 1961

---

4. Senator Glenn, when you became the first American to orbit the Earth aboard Friendship 7 in 1962, you had the

opportunity to see our planet from space for the first time. What was that experience like, and what impression did it leave on you?

Seeing Earth from orbit for the first time was one of the most extraordinary experiences of my life. As **Friendship 7** traveled around the planet, I had the opportunity to look through the spacecraft's window and observe the Earth below. What struck me immediately was the incredible beauty of the planet. The oceans appeared as deep shades of blue, the clouds formed vast swirling patterns across the atmosphere, and the continents stretched out in remarkable detail.

One of the most striking things was the **thinness of the atmosphere**. From orbit, it appeared as a delicate blue line surrounding the Earth. It made me realize how fragile that protective layer is and how important it is for sustaining life on our planet.

Another powerful impression was how quickly the landscape changed as we moved around the Earth. Within minutes, we passed from daylight into darkness, watching the Sun rise and set in ways that are impossible to experience on the ground.

Perhaps the most important realization was that from space **you cannot see national borders**. The divisions that sometimes separate people on Earth simply disappear from that vantage point. Instead, you see one beautiful world shared by all humanity.

That perspective stays with you long after the flight ends. It reminds you how precious our planet is and how important it is for people to work together to protect it. Spaceflight changes the way you see the Earth—not just as a place where we live, but as a remarkable home for all humankind.

---

5. Senator Glenn, during your orbital mission aboard Friendship 7 in 1962, mission control became concerned about a possible problem with the spacecraft's heat shield. Could you describe what happened, how the issue was handled during the flight, and what was going through your mind at the time?

Yes, that was one of the more tense moments during the mission. During my third orbit around the Earth, the flight controllers at NASA noticed a signal from one of the spacecraft's sensors that suggested the **landing bag**—a device located beneath the heat shield—might have deployed prematurely. If that had happened, it could have meant that the **heat shield protecting the capsule during reentry might be loose or compromised**.

The heat shield was critical. When the spacecraft reentered Earth's atmosphere, temperatures outside the capsule could reach several thousand degrees. Without a properly secured heat shield, the spacecraft would not survive reentry.

Because there was uncertainty about the sensor reading, mission control decided to take a precautionary approach. Normally, the **retro-rocket package**, the small rockets used to slow the spacecraft for reentry, would be jettisoned after firing. Instead, the flight controllers instructed me to **leave the retro-rocket pack attached to the capsule during reentry**. The idea was that the straps holding the retro package might help keep the heat shield in place if it had loosened.

Inside the spacecraft, I was aware that something unusual was happening, although the details were not fully clear to me during the flight. I knew the retro pack had been left in place and that we would simply follow the procedures and continue with the reentry sequence.

As I descended through the atmosphere, I could see **pieces of burning debris passing by the window**, which were likely fragments of the retro-rocket package burning away as the spacecraft slowed down. It was a dramatic sight, but the capsule performed exactly as designed.

Fortunately, after landing it was discovered that the **heat shield had never actually been loose**. The warning had been caused by a faulty sensor reading. In the end, the spacecraft functioned perfectly during reentry, and the mission concluded safely.

That situation illustrated something important about the early days of spaceflight: every mission involved unknown, and both astronauts and engineers had to be prepared to adapt to unexpected situations. The teamwork between the astronaut in the capsule and the engineers on the ground played a crucial role in the success of the mission. 🚀

Glenn in his Mercury spacesuit in 1962

---

6. Senator Glenn, during the early days of the Mercury program there was a story that astronauts carried a small note during their missions explaining who they were and why they had arrived. The note reportedly read: *"I am a stranger. I come in peace. Take me to your leader and there will be a massive reward for you in eternity,"* and was written in several languages in case the capsule landed

near remote islands in the Pacific Ocean. Was this your idea, or was it something NASA prepared?

The early Mercury missions involved a great deal of uncertainty. Our spacecraft were relatively small, and although recovery operations were carefully planned, there was always a possibility that a capsule might land in a remote area far from the expected recovery zone.

Because many of the splashdown areas were in the **Pacific Ocean**, there was a chance that an astronaut might land near small islands where the local population had never encountered a spacecraft before. NASA and the mission planners tried to think ahead about how to handle such unusual situations.

As part of that preparation, small **identification cards and notes** were created that astronauts could use if they found themselves in an unexpected location. These notes were written in several major languages and explained that the person arriving in the spacecraft came in peace and needed assistance contacting authorities.

It was not really a personal idea of mine but rather a practical precaution developed by NASA and the mission planners. In those early days of spaceflight, we tried to anticipate as many scenarios as possible—no matter how unlikely they might seem. Fortunately, the Mercury missions generally landed very close to the planned recovery areas, where U.S. Navy ships and helicopters were already waiting. As a result, the notes were never actually needed.

Still, the story reflects something about the early space program: we were venturing into a completely new frontier, and everyone involved—from astronauts to engineers—was preparing for every possible outcome.

---

7. Colonel Glenn, in the early 1960s NASA considered the possibility of recruiting women into the astronaut corps, and a group of highly qualified female pilots later became known as the "Mercury 13." During hearings before the United States House Committee on Science and Astronautics in 1962, you expressed reservations about sending women into space at that time. Could you explain why you held that view then, and how you reflect on that issue today?

It's important to understand the context of the early space program in the early 1960s. At that time, NASA's astronaut selection criteria were based largely on the backgrounds of military test pilots. The reasoning was that these individuals already had extensive experience flying high-performance aircraft, managing complex systems under pressure, and operating in dangerous experimental conditions.

In those years, military test pilot positions in the United States were not open to women, which meant that the pool of candidates meeting NASA's criteria consisted entirely of men. My comments during the congressional hearings reflected that reality. I explained that the astronaut corps was drawn from that specific group of experienced test pilots because the space program was still very experimental and extremely risky.

Looking back, I believe my remarks were describing how the system worked at the time rather than expressing a personal belief about women's abilities. The social and professional opportunities available to women in aviation were very different then than they are today.

Over the following decades, those barriers began to change. When NASA expanded its astronaut corps for the Space Shuttle program, the agency opened the selection process to scientists, engineers, and pilots from a much broader range of backgrounds.

As a result, outstanding astronauts such as Sally Ride and many others demonstrated beyond any doubt that women could perform every aspect of spaceflight with great skill and professionalism. Space exploration has always evolved alongside society. As opportunities expanded, the astronaut corps became more diverse, reflecting the talents and contributions of people from many backgrounds.

Today it is clear that the exploration of space belongs to all of humanity, and the achievements of women astronauts have become an essential and inspiring part of that story.

---

8. Senator Glenn, your orbital flight aboard Friendship 7 in 1962 made you the first American to orbit the Earth. How did that historic mission influence public perception of space exploration in the United States and around the world?

The flight had a significant impact on how people viewed space exploration. At the time, the United States was still in the early stages of the space program, and many people were watching closely to see whether humans could successfully travel into orbit and return safely. My mission demonstrated that this was possible, and it helped build confidence in the nation's growing space capabilities.

The public response was extraordinary. When I returned to Earth, millions of people followed the news of the mission, and there was a tremendous sense of pride and excitement. Shortly afterward, there was a large parade in New York City, and I was honored to meet many Americans who felt that the mission represented an important step forward for science and exploration.

Beyond the United States, the mission also contributed to the global excitement surrounding the Space Age. People around the world were fascinated by the idea that humans could leave Earth and view our planet from space.

Perhaps most importantly, the early Mercury flights helped inspire young people to study science, engineering, and mathematics. Many students who watched those missions later became scientists, engineers, and astronauts themselves.

Space exploration has always had the power to capture imagination. The Mercury program showed that the dream of traveling beyond Earth was no longer science fiction, it had become a reality. And that realization encouraged many people to believe that even greater discoveries were possible in the future.

---

9. Senator Glenn, many astronauts have described spaceflight as a life-changing experience, particularly when they see Earth from orbit. Did your journey into space change the way you viewed our planet and humanity's place on it?

Yes, it certainly did. When I orbited the Earth aboard **Friendship 7** in 1962, one of the most powerful impressions was seeing our planet from space for the first time. From orbit, Earth appears as a beautiful blue sphere surrounded by the darkness of space. The atmosphere, which sustains all life, looks incredibly thin and fragile.

From that perspective, national borders disappear. You don't see political divisions or conflicts—you simply see one planet shared by all humanity. It creates a strong sense that we are all connected and that we share responsibility for protecting our world. I was also struck by the incredible beauty of the Earth. The oceans, clouds, and continents form a constantly changing landscape that is both majestic and delicate.

Many astronauts have spoken about what is sometimes called the **"overview effect,"** the realization that our planet is small compared to the vastness of space. Experiencing that firsthand encourages a deeper appreciation for the environment and for the cooperation needed to address global challenges.

For me, spaceflight reinforced the idea that exploration has great value—not only for advancing science and technology, but also for helping humanity see itself from a broader perspective. When you look at Earth from space, you realize that our differences are small compared to what we share as inhabitants of this remarkable planet. 🚀

---

10. Senator Glenn, after your distinguished career as an astronaut and Marine Corps pilot, you entered public service and even announced your candidacy for the presidency of the United States in 1983. Could you elaborate on your political career and what motivated you to pursue leadership in government?

After my years in the space program and in military service, I felt a strong desire to continue contributing to the country in another capacity—through public service. My experiences as a Marine Corps pilot, astronaut, and participant in the early space program gave me a unique perspective on national priorities such as science, technology, education, and national security. I believed that those areas were critical to the future of the United States, and I wanted to help shape policies that would strengthen them.

In 1974, I was elected to the United States Senate from Ohio and served there for four consecutive terms, representing the state for twenty-four years. During my time in the Senate, I worked on a variety of issues, including arms control, scientific research, education, and government accountability. I also chaired the United States Senate Governmental Affairs Committee, where we focused on improving the efficiency and transparency of federal programs.

In 1983, I announced my candidacy for the presidency in the 1984 election. My campaign focused on strengthening the economy, supporting technological innovation, and maintaining a strong national defense while promoting international cooperation. Although the campaign did not ultimately succeed, it was an important experience. Running for president allowed me to engage directly with people across the country and better understand the challenges facing American communities.

Public service is always a team effort, and I was proud to represent Ohio and work with colleagues from both parties in the Senate. One of the most meaningful moments of my later career came in 1998 when I returned to space aboard the **Space Shuttle Discovery**. That flight allowed scientists to study the effects of aging in space and symbolized how exploration can continue throughout a lifetime.

Looking back, I consider my political career to be a natural extension of the same spirit that drove the early space program, the desire to serve the public and help build a better future.

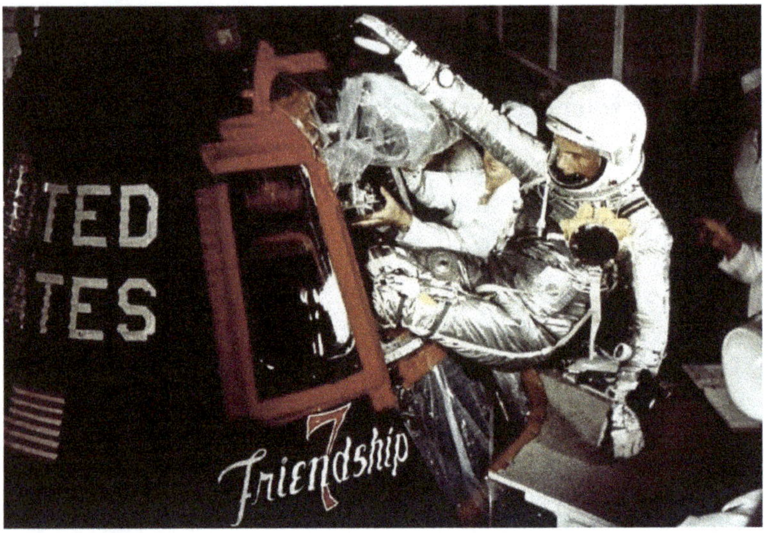

Glenn entering his spacecraft, Friendship 7, prior to the launch of Mercury-Atlas 6, 1962

---

11. Decades later, you returned to space aboard the Space Shuttle at age seventy-seven. What motivated you to fly again?

The mission allowed scientists to study the effects of spaceflight on the aging human body. I believed that participating in that research could contribute valuable knowledge that would help future astronauts.

12. What do you believe is the greatest lesson humanity has learned from space exploration?

One of the greatest lessons is that exploration expands knowledge and brings people together. Space exploration has the power to inspire cooperation, innovation, and curiosity about the universe.

Glenn presents President Kennedy with an American flag he carried inside his space suit on Friendship 7.

13. Senator Glenn, during your long career in public service you had the opportunity to meet many presidents, international leaders, and influential figures around the world. Among the people you encountered, which

individuals impressed you the most, and what qualities made them stand out for you?

Over the years I had the privilege of meeting many remarkable leaders, both in the United States and internationally. What impressed me most about the individuals who stood out was not only their authority or position, but their character, judgment, and sense of responsibility.

One leader who left a strong impression on me was **John F. Kennedy**. His decision in 1961 to challenge the nation to land a man on the Moon before the end of the decade was bold and visionary. At the time, it seemed almost impossible, yet that vision inspired an entire generation of scientists, engineers, and astronauts to accomplish something extraordinary.

I was also impressed by **Ronald Reagan**, whom I interacted with during my time in the Senate. President Reagan had a remarkable ability to communicate with the American people and to express complex ideas in ways that were clear and inspiring.

On the international stage, leaders who emphasized cooperation and dialogue always stood out to me. For example, **Mikhail Gorbachev** played a crucial role in reducing tensions during the final years of the Cold War. His willingness to pursue reforms and engage with the international community helped create an environment where the world could move toward greater stability.

What I found most impressive about the best leaders I met was their ability to think beyond short-term political interests and focus on the long-term well-being of their countries and the world. Leadership, whether in government, science, or exploration, ultimately comes down to vision, integrity, and the ability to bring people together to accomplish important goals. Those are the qualities that leave the most lasting impact.

---

14. Senator Glenn, you were one of the pioneers of human spaceflight and later returned to space many decades later. What message would you share with future generations of astronauts who will explore the Moon, Mars, and beyond?

To future astronauts, I would say that you are continuing one of humanity's most inspiring journeys. When we flew the early missions of the **Mercury program**, including my flight aboard **Friendship 7**, we were taking the very first steps into space. Much of what we were doing had never been attempted before, and every mission helped humanity learn how to travel beyond Earth safely.

The astronauts of today and tomorrow will build upon that foundation and go much farther than we ever did. Exploration requires courage, preparation, and teamwork. Every astronaut represents not only their own abilities but also the efforts of thousands of scientists, engineers, and technicians who make space missions possible. Never forget that you are part of a much larger team working together to expand human knowledge.

I would also encourage future explorers to remember the perspective that spaceflight provides. When you see Earth from space, you realize how beautiful and fragile our planet is. That view reminds us that humanity shares one home and one future.

The missions ahead—whether returning to the Moon or traveling to Mars—will require creativity, determination, and international cooperation. They will also inspire new generations of students to study science, engineering, and exploration.

Every great journey begins with curiosity. So, remain curious, remain committed to discovery, and continue pushing the boundaries of what humanity can achieve. The future of exploration belongs to those who are willing to reach the stars.

Glenn on the Space Shuttle

Interview with Gus Grissom

Pioneer of the Mercury and Gemini Programs

Introduction

In the earliest years of the Space Age, astronauts were not only explorers, but they were also test pilots of machines that had never been proven. Every launch carried uncertainty. Every mission pushed the limits of engineering, physics, and human endurance. Yet a group of courageous individuals volunteered to climb aboard rockets and venture into space for the very first time.

Among them was Virgil "Gus" Grissom. Born in Indiana and raised with the determination typical of many early astronauts, Grissom became a skilled fighter pilot and later an exceptional test pilot. His calm judgment and technical precision made him a natural candidate when NASA selected its first group of astronauts, the legendary ***Mercury Seven****.*

In 1961, Grissom piloted the ***Liberty Bell 7*** *spacecraft on the second American human spaceflight mission. Like Alan Shepard's earlier flight, it was a brief suborbital journey, but it helped prove that humans could travel safely into space. The mission ended dramatically when the spacecraft's hatch* *unexpectedly opened after splashing down in the Atlantic Ocean, causing the capsule to sink. Although Grissom escaped safely, the incident became one of the most debated moments of the early space program.*

Yet NASA's confidence in Grissom never wavered. He went on to command the first crewed ***Gemini mission****, helping demonstrate techniques that would later make lunar exploration possible. His experience, leadership, and technical skill made him one of NASA's most respected astronauts.*

In 1967, Grissom was chosen to command the first crewed mission of the Apollo program—an ambitious effort that would eventually land humans on the Moon. But during a ground test for ***Apollo 1****, tragedy struck when a cabin fire claimed the lives of Grissom and his fellow astronauts Roger Chaffee and Ed White.*

Their sacrifice deeply shook NASA but ultimately led to major improvements in spacecraft safety—changes that made the later Apollo missions possible. Today, Gus Grissom is remembered not only as a skilled astronaut but also as one of the pioneers who helped lay the foundation for humanity's journey to the Moon. His story is one of courage, dedication, and the willingness to explore the unknown.

Virgil Ivan "Gus" Grissom (April 3, 1926 – January 27, 1967) was an American engineer and pilot in the United States Air Force, as well as one of the original Mercury Seven selected by the National Aeronautics and Space Administration for Project Mercury, a program to train and launch astronauts into outer space. Grissom went on to be a Project Gemini and Apollo program astronaut for NASA. As a member of the NASA Astronaut Corps, Grissom was the second American to fly in space in 1961. He was also the second American to fly in space twice, preceded only by Joe Walker with his sub-orbital X-15 flights.

---

1. Commander Grissom, what first inspired you to become a pilot?

Growing up in Indiana, aviation seemed like an exciting and challenging field. During World War II and the years that followed, aviation technology advanced rapidly, and I wanted to be part of that progress. Flying gave me the opportunity to push myself and to explore the limits of what aircraft could do.

---

2. What did it mean to you to be selected as one of NASA's original Mercury astronauts?

It was an incredible opportunity and a tremendous responsibility. The Mercury program represented the first attempt by the United States to send humans into space. All of us knew that the missions would involve risk, but we also knew that we were participating in something historic.

Grissom (far left) with fellow Project Mercury astronauts and a model of the Atlas rocket, July 12, 1962

---

3. Major Grissom, your mission aboard Liberty Bell 7 in 1961 made you the second American to travel into space. What do you remember most clearly about the launch and the moments as the rocket lifted you toward space?

The launch of Liberty Bell 7 was an unforgettable experience. As the countdown reached zero and the **Redstone rocket** engines ignited, the spacecraft began to vibrate as the rocket built up thrust. A moment later we lifted off the launch pad, and I could feel the strong acceleration pushing me firmly back into the seat.

Compared with what many people might imagine the launch itself was actually very smooth. The rocket climbed steadily and predictably, and the spacecraft systems were performing just as they had during simulations.

During ascent the acceleration increased as the rocket burned fuel, and you could clearly feel the forces pressing against your body. At the same time, I was carefully monitoring the spacecraft's instruments and communications with ground control.

After a few minutes the engine shut down and suddenly everything became quiet. That was the moment when weightlessness began. Objects inside the capsule started to float, and it was a remarkable sensation to realize that we were truly in space.

Looking out through the small window, I could see the curvature of the Earth and the deep blackness of space beyond it. Even though the mission lasted only about fifteen minutes, those moments provided a powerful glimpse of what human spaceflight could become. The flight demonstrated that American astronauts could successfully travel into space and return safely, which helped pave the way for the longer orbital missions that followed soon afterward.

---

4. How did it feel to experience weightlessness for the first time?

Weightlessness was a fascinating sensation. Everything inside the spacecraft floated freely, and you had to adapt quickly to the new environment. It confirmed that humans could function effectively in space, which was an important discovery for future missions.

Grissom in front of the Liberty Bell 7 spacecraft

---

5. Major Grissom, your flight aboard Liberty Bell 7 ended with a dramatic moment after splashdown when the

capsule's hatch unexpectedly opened and the spacecraft quickly filled with water. What do you remember about that moment, and what do you believe caused the hatch to open?

The mission itself went very well, and the spacecraft performed exactly as expected during launch, flight, and reentry. After reentering the atmosphere, the capsule deployed its parachutes normally and splashed down in the Atlantic Ocean.

Once we were in the water, I began going through the normal post-landing procedures. Recovery helicopters from the **United States Navy** were already approaching the capsule.

Suddenly, without warning, the explosive side hatch blew open. The Mercury spacecraft was designed with a hatch that could be opened quickly using small explosive bolts. This system was intended to allow astronauts to exit the capsule rapidly in an emergency. However, when the hatch opened unexpectedly, seawater immediately began rushing into the spacecraft.

I realized the capsule was filling with water and began exiting the spacecraft as quickly as possible. Once outside, I swam away from the capsule and waited for the recovery helicopter to pick me up.

Meanwhile, the Liberty Bell 7 capsule quickly became too heavy with water and eventually sank into the ocean before it could be secured by the recovery team.

For many years there was debate about what caused the hatch to open. Some people initially wondered whether I might have accidentally triggered the hatch release. However, I have always maintained that **I** did not activate the hatch. Later investigations suggested that the hatch may have opened due to a mechanical or electrical malfunction, although the exact cause was never definitively determined.

The important thing is that the mission itself demonstrated that American astronauts could safely launch into space and return to Earth. Even though the capsule was lost, the flight provided valuable experience that helped improve procedures for later Mercury missions. 🚀

---

6. Major Grissom, after your Mercury mission aboard Liberty Bell 7, you went on to command Gemini 3 mission. How did Project Gemini differ from the earlier Project Mercury, and what new capabilities did it introduce?

The Gemini program represented a major step forward from the Mercury missions. Mercury was designed primarily to answer one basic question: Could a human survive and function in space? The spacecraft carried only one astronaut and had relatively limited capabilities. During those flights we demonstrated that humans could travel into orbit, experience weightlessness, and return safely to Earth.

Gemini, on the other hand, was designed to prepare NASA for the Apollo missions and the goal of landing astronauts on the Moon. One of the biggest differences was that the Gemini spacecraft carried two astronauts instead of one, which allowed for better teamwork and more complex mission operations. During Gemini 3, my crewmate **John Young** and I tested the spacecraft's maneuvering system.

Unlike Mercury, which had very limited ability to change its position in orbit, Gemini spacecraft could maneuver using thrusters. This capability was extremely important because future missions would require astronauts to rendezvous and dock with another spacecraft in orbit.

Gemini also introduced longer-duration flights, spacewalks, and precision reentry control. These techniques were essential for the Apollo program. Landing on the Moon required astronauts to dock spacecraft in orbit, perform complex navigation, and work outside the spacecraft in space.

In many ways, Gemini was the program where astronauts and engineers learned how to operate effectively in space, not just survive there. Mercury opened the door to space, but Gemini taught us how to work there—and those lessons made the Apollo missions possible. 🚀

---

7. What do you believe was the most important achievement of the Gemini missions?

Gemini demonstrated that astronauts could operate spacecraft with much greater control and flexibility. The program developed the techniques needed for lunar missions, including orbital rendezvous, which later became a key part of the Apollo program.

Grissom with the Apollo 1 crew in 1966

---

8. Major Grissom, the early days of human spaceflight involved significant uncertainty and risk. What motivated astronauts like you to accept those dangers and participate in the pioneering missions of the Space Age?

Space exploration in the early 1960s was indeed risky, and everyone involved understood that. But for many of us, the motivation came from a combination of curiosity, professional challenge, and a sense of responsibility to push the boundaries of human knowledge.

Most of the early astronauts were test pilots before joining NASA. As test pilots, we were already accustomed to flying experimental aircraft and evaluating new technologies. Accepting a certain level of risk was part of that profession. When the opportunity arose to help develop spacecraft capable of traveling beyond Earth, it felt like a natural extension of what we had been doing.

Another powerful motivation was the idea of exploration. Humanity has always been driven to explore new frontiers—from crossing oceans to reaching the highest mountains. Space represented the next great frontier.

Programs such as **Project Mercury** and later **Project Gemini** were about more than individual missions. They were about learning how humans could live and work in space so that future missions could go farther.

We also believed strongly in the team behind us. Thousands of engineers, scientists, and technicians worked tirelessly to design and test the spacecraft. Their dedication gave us confidence that the risks were being carefully managed.

In the end, what motivated us most was the belief that what we were doing would open the door for future generations. The early missions were only the beginning, and we knew that someday astronauts would travel much farther—perhaps to the Moon, Mars, and beyond. Being part of that first step was an extraordinary privilege. 🚀

---

9. Major Grissom, you were one of the pioneers of human spaceflight during the Project Mercury and later commanded the Gemini 3 mission. How do you hope future generations will remember the early astronauts who helped launch the Space Age?

I hope future generations remember the early astronauts not simply as individuals, but as part of a much larger team that helped open the door to space. In the early days of the space program, everything we were doing was new. Engineers were designing rockets and spacecraft that had never existed before, and astronauts were learning how to operate in an environment no human had experienced.

Programs like **Project Mercury** and **Project Gemini** were not just about flying into space. They were about building the knowledge and technology needed for the next great step—landing humans on the Moon through the **Apollo program**.

The early astronauts understood that we were part of an experiment in exploration. Every mission taught us something new about spacecraft, about the environment of space, and about how humans could work there safely. But the real achievement belonged to the thousands of people behind the scenes—the engineers, technicians, scientists, and mission controllers who made every flight possible.

If people remember us at all, I hope they see us as pioneers who helped prove that humans could leave Earth and explore beyond it. Those first steps were only the beginning. The real story of space exploration will be written by the generations that follow, as they travel farther into the solar system and continue expanding humanity's understanding of the universe. 🚀

---

10. Major Grissom, as one of the pioneers of the early Space Age who flew during the Project Mercury and later commanded the Gemini 3 mission, what message would you share with future explorers who will travel beyond Earth to the Moon, Mars, and perhaps even farther?

To the explorers who will travel beyond Earth in the future, I would say that you are continuing one of humanity's greatest adventures. When we flew the early missions of the Mercury and Gemini programs, we were taking the very first steps into space. Much of what we were doing had never been attempted before. Every flight helped engineers and astronauts learn something new about how humans could survive and operate in space.

The missions that will take people to the Moon again, and eventually to Mars, will be even more complex and demanding. Those journeys will require advanced technology, careful preparation, and strong teamwork.

My advice to future explorers is to respect the challenges of spaceflight and trust the people who work alongside you. No astronaut ever reaches space alone. Every mission depends on the efforts of thousands of scientists, engineers, technicians, and mission controllers who dedicate their lives to making exploration possible.

You should also remember that exploration is not only about technology, it is about curiosity and courage. Throughout history, humanity has always moved forward by exploring new frontiers, whether crossing oceans or reaching space.

The early astronauts helped open the door to the universe. The explorers of the future will carry that journey much farther than we ever could. So be prepared, be disciplined, and never lose the sense of wonder that drives people to look up at the stars and ask what lies beyond. 🚀

Interview with Alexei Leonov

The First Human to Walk in Space

Introduction

By the mid-1960s, humanity had already taken its first steps beyond Earth. Astronauts and cosmonauts had traveled into orbit and returned safely. Yet space remained an environment largely unexplored by human hands. One question remained unanswered: Could a human leave the safety of a spacecraft and work directly in the vacuum of space?

On March 18, 1965, Soviet cosmonaut Alexei Leonov prepared to answer that question. Aboard the spacecraft ***Voskhod 2****, Leonov and his commander Pavel Belyayev orbited the Earth as engineers and mission controllers monitored every detail of the historic flight. Attached to the spacecraft was an inflatable airlock designed to allow a cosmonaut to exit the vehicle while maintaining cabin pressure.*

No human had ever attempted such a maneuver. When the moment arrived, Leonov opened the hatch and slowly pushed himself into the darkness of space. For the first time in history, a human being floated freely outside a spacecraft.

Suspended above Earth and connected only by a tether, Leonov drifted through the silence of space. Beneath him stretched the vast curvature of the planet—oceans, clouds, and continents moving slowly across the horizon. The experience was breathtaking. But it was also dangerous.

The vacuum of space caused Leonov's spacesuit to expand, becoming so rigid that he struggled to bend his arms and legs. When it was time to return to the spacecraft, he realized he could barely fit back through the airlock.

With remarkable calm and courage, Leonov partially released pressure from his suit—a risky maneuver—and managed to squeeze back into the spacecraft, narrowly avoiding disaster. His spacewalk lasted only about twelve minutes, but it transformed the future of human spaceflight.

Today, spacewalks—known as ***extravehicular activities (EVAs)****—are essential for maintaining spacecraft, assembling space stations, and conducting scientific research in orbit.*

Alexei Leonov's historic step outside his spacecraft opened a new dimension of exploration. He later became commander of the Soviet spacecraft in the historic ***Apollo–Soyuz Test Project****, the first joint space mission between the United States and the Soviet Union—a mission that symbolized cooperation in a frontier once defined by rivalry.*

Leonov was not only a cosmonaut but also an artist, capturing scenes of space exploration in paintings inspired by his experiences. His courage and imagination helped humanity move closer to mastering life beyond Earth.

Alexei Arkhipovich Leonov (30 May 1934 – 11 October 2019) was a Soviet and Russian cosmonaut and aviator, Air Force major general, writer, and artist. On 18 March 1965, he became the first person to conduct a spacewalk, exiting the capsule during the Voskhod 2 mission for 12 minutes and 9 seconds. He was also selected to be the first Soviet person to land on the Moon although the project was eventually cancelled.

1. Alexei Arkhipovich, when did your dream of flying and exploration begin?

My fascination with flight began during my youth when aviation captured the imagination of many people in the Soviet Union. I loved drawing airplanes and studying how they worked. Eventually, I became a military pilot, which led me toward the opportunity to join the cosmonaut program.

---

2. What was your reaction when you were selected to participate in the Voskhod program?

It was an extraordinary honor. At that time, every mission represented a new experiment in human spaceflight. We understood that we were pioneers entering an environment that was still largely unknown.

Alexei Leonov (far left, back row) with fellow cosmonauts in 1965 Photographed by Alexander Mokletsov, Source: Wikimedia Commons CC BY-SA 3.0

3. Colonel Leonov, during the historic Voskhod 2 mission in 1965, you became the first human to leave a spacecraft and walk in open space. Before attempting this unprecedented spacewalk, did you feel nervous or apprehensive about what might happen?

Of course there was a sense of tension before the spacewalk, but it was not the kind of fear that prevents you from doing your job. It was more a feeling of deep concentration and responsibility.

At that time, no human had ever left the safety of a spacecraft and floated freely in space. We did not fully know how the human body would react, how the spacesuit would behave in vacuum, or how easily I would be able to move and return to the spacecraft. But we trained extensively for the mission.

The engineers and designers had worked very hard to prepare the **Voskhod 2 spacecraft**, and my crewmate **Pavel Belyayev** and I had practiced the procedures many times on Earth. That preparation helped give me confidence.

When the airlock opened and I moved outside the spacecraft, I was immediately struck by the incredible view. The Earth beneath me was magnificent—bright blue oceans, white clouds, and the thin atmosphere glowing against the darkness of space.

In that moment, any nervousness disappeared. Instead, I felt a sense of wonder and pride that humanity had taken another step into the cosmos. The experience was both technically challenging and emotionally powerful. It reminded me that exploration always involves uncertainty, but it is through those moments of courage that humanity advances into new frontiers. 🚀

---

4. Colonel Leonov, during the historic Voskhod 2 mission, you became the first person in history to float freely outside a spacecraft. What was that moment like when you first emerged from the airlock into open space?

The moment I left the spacecraft was one of the most extraordinary experiences of my life. When the airlock opened and I began to move outside the **Voskhod 2 spacecraft**, I suddenly found myself floating above the Earth with nothing but the safety line connecting me to the spacecraft. It was a completely new sensation, there was no sense of up or down, only the vastness of space around me.

The view was breathtaking. Below me, the Earth appeared as a magnificent blue sphere covered with swirling white clouds. The sunlight reflected off the oceans and atmosphere, creating colors that are difficult to describe. Above me was the deep blackness of space filled with brilliant stars.

I felt both very small and incredibly privileged at that moment. For the first time in history, a human being was standing—or rather floating—outside a spacecraft in open space. I realized that this was not only a personal achievement but a milestone for all humanity.

At the same time, I had to remain focused on my tasks. Spacewalking required careful movement and constant attention to the equipment and safety procedures. Still, I remember taking a moment simply to look at the Earth and appreciate the beauty of our planet. That image has remained with me ever since. It was a moment that symbolized humanity taking another step into the vast frontier of space. 🚀

March 1965, the first space walk

---

5. Colonel Leonov, during the historic Voskhod 2 mission, your spacesuit unexpectedly expanded in the vacuum of space, making it extremely difficult for you to return to

the airlock. What was going through your mind during that dangerous moment?

Yes, that was one of the most challenging moments of the mission. When I was outside the **Voskhod 2 spacecraft**, I soon realized that my spacesuit had expanded more than expected. In the vacuum of space, the suit became rigid and inflated, which made it very difficult for me to bend my arms and legs or maneuver normally.

When it came time to return to the airlock, I discovered that I could not easily fit back inside. The suit had become so stiff that I could not enter the airlock in the normal way. At that moment, I understood that I had to remain calm and think carefully. Panic would only make the situation worse.

I made the decision to reduce the pressure inside my spacesuit slightly, even though that was a risky action. Lowering the pressure allowed the suit to become more flexible so that I could move my body enough to enter the airlock.

Even after that, I had to enter the airlock head first instead of feet first, which was not the original procedure. Once inside, I slowly repositioned myself so the outer hatch could be closed. The entire situation required careful thinking and control. Space exploration often involves unexpected challenges, and astronauts must be prepared to solve problems under pressure.

When I finally reentered the spacecraft and closed the hatch, I felt enormous relief. We had successfully completed the first human spacewalk, and despite the difficulties, the mission had achieved its goal. That experience demonstrated how important preparation, discipline, and calm decision-making are when exploring new frontiers. 🚀

The Apollo-Soyuz crew in 1975

---

6. Colonel Leonov, after completing the first human spacewalk during the Voskhod 2 mission, how did that experience change your understanding of space exploration and the challenges astronauts face in the future?

The experience taught me that space exploration is both more difficult and more remarkable than we had imagined.

Before the mission, engineers and scientists had carefully planned every detail of the spacewalk. We conducted extensive training and simulations on Earth, but space is an environment that always contains surprises. When my spacesuit expanded in the vacuum, it reminded us that even the best preparation cannot predict every situation.

That moment helped me understand that flexibility and problem-solving are essential qualities for astronauts. In space, you must remain calm and think clearly when unexpected challenges arise. At the same time, the experience also showed me the enormous potential of human exploration.

Floating outside the spacecraft and seeing the Earth beneath me was a powerful reminder that humanity had taken a new step into the universe. For the first time, a human being was working outside a spacecraft in open space. That capability would later become essential for building space stations, repairing satellites, and conducting scientific research in orbit.

The first spacewalk demonstrated that humans could operate not only inside spacecraft but also in the harsh environment of space itself. It also reinforced another important lesson: space exploration is never the work of a single person. My success depended on the engineers who designed the spacecraft and spacesuit, the training teams who prepared us, and my crewmate **Pavel Belyayev**, who remained inside the spacecraft ready to assist if necessary.

That experience made me appreciate both the complexity and the promise of space exploration. Each mission teaches us something new, and each challenge helps humanity move closer to exploring the wider universe. 🚀

---

7. Colonel Leonov, years after your historic spacewalk, you served as commander of the Soviet crew during the Apollo–Soyuz Test Project, the first cooperative space mission between the United States and the Soviet Union. What did that moment of international cooperation mean to you personally?

The **Apollo–Soyuz mission** was a very special moment in the history of space exploration. During the early years of the Space Age, the Soviet Union and the United States were competitors. Both nations were striving to achieve important milestones in space, and that rivalry helped accelerate technological progress. However, space itself belongs to all humanity, and eventually we began to recognize the importance of cooperation.

In 1975, when our **Soyuz spacecraft** docked with the American **Apollo spacecraft**, it symbolized something very important: that even nations with different political systems could work together in the peaceful exploration of space.

When the docking was completed and I opened the hatch to greet **Thomas P. Stafford**, we shook hands in orbit. That handshake became a symbol around the world that cooperation could replace competition. For me personally, the mission demonstrated that astronauts and cosmonauts share the same goals. We all want to explore the universe, expand scientific knowledge, and contribute to humanity's progress.

The Apollo–Soyuz mission also helped establish a tradition of international collaboration that continues today in projects such as the **International Space Station**, where astronauts from many countries live and work together. To me, that was the most meaningful lesson of the mission: space exploration should unite humanity rather than divide it. When you look at Earth from space, you do not see borders, you see one beautiful planet that belongs to all of us. 🚀

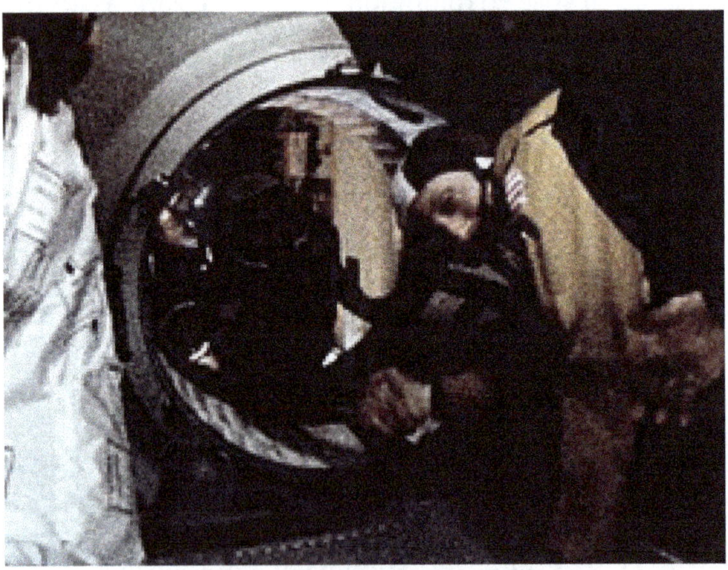

The historic handshake between Leonov and Tom Stafford

---

8. Colonel Leonov, in addition to being a pioneering cosmonaut, you are also known as a talented artist. Did your experiences in space influence your artwork, and how did space exploration inspire your paintings?

Yes, space had a profound influence on my artwork. Even before becoming a cosmonaut, I had a strong interest in drawing and painting. Art was always a way for me to express what I observed and felt. When I traveled into space aboard the **Voskhod 2 mission**, that artistic perspective became even more important.

The view of Earth from space is something very difficult to describe with words alone. The colors, the vastness of the planet, and the contrast between the bright blue Earth and the deep blackness of space create an image that is both beautiful and emotional.

When I performed the first spacewalk from the **Voskhod 2 spacecraft**, I had the opportunity to see our planet from a perspective that very few people had experienced. I remember looking at the horizon where the atmosphere glowed with a thin blue light, and the clouds cast delicate shadows over the oceans.

Those images stayed with me. After returning from space, I began creating paintings that attempted to capture those moments. Photography can record what space looks like, but art can convey the feeling of being there, the sense of wonder, the silence of space, and the beauty of our planet.

Later in my career, I also created artwork illustrating historic missions such as the **Apollo–Soyuz Test Project**, which symbolized cooperation between nations in space exploration.

For me, art became another way to share the experience of space with people on Earth. Through painting, I hoped to help others imagine what it feels like to float above our planet and look out into the infinite universe. In that sense, space exploration did not only inspire scientific discovery, it also inspired creativity and a deeper appreciation for the beauty of our world. 🚀🎨

Leonov's 1967 painting Near the Moon and a still from 2001: A Space Odyssey, 1968

9. What do you believe is the greatest lesson humanity has learned from space exploration?

One of the most important lessons of space exploration is how small and fragile our planet truly is. When you look at Earth from space, you see a beautiful blue sphere surrounded by the vast darkness of the universe. The atmosphere that protects life appears to be only a thin, delicate layer. From that perspective, the divisions between countries disappear. There are no borders, no political lines, only one planet shared by all humanity.

That view reminds us that we all live on the same world and share the same responsibility to protect it. Space exploration not only expands our knowledge of the universe; it also teaches us an important lesson about cooperation and stewardship of our home. When astronauts look back at Earth, they often realize that humanity's greatest challenge is not exploring distant worlds but working together to care for the one we already have. 🌍 🚀

The Apollo-Soyuz crew in 1975

10. Colonel Leonov, as one of the pioneers of human spaceflight and the first person to walk in open space during the Voskhod 2 mission, what message would you share with future explorers who will travel deeper into space—to the Moon, Mars, and beyond?

To the explorers of the future, I would say that you are continuing one of humanity's most important journeys. When we flew the early missions of the Space Age, we were taking the very first steps beyond Earth. At that time, even leaving the spacecraft and floating in space was something completely unknown. Yet those early missions proved that human beings could work and survive in the environment of space.

The explorers who will travel to the Moon again and eventually to Mars will face challenges far greater than those we experienced. These missions will require advanced technology, long-duration spaceflight, and extraordinary cooperation between nations.

My advice to future astronauts is to remain curious, disciplined, and prepared. Space exploration demands not only courage but also careful training and teamwork. Every mission represents the efforts of thousands of scientists, engineers, and technicians who make exploration possible.

I would also remind future explorers of an important lesson that space teaches us. When you look back at Earth from space, you see a small, beautiful planet without borders. That perspective reminds us that humanity shares one home and one destiny.

For this reason, I believe that the exploration of space should always be a peaceful endeavor that unites nations rather than divides them. The universe is vast, and humanity has only begun to explore it. The explorers of the future will travel farther than we ever imagined, discovering new worlds and expanding our understanding of the cosmos. Carry that journey forward with courage, cooperation, and respect for the fragile planet from which you began. 🚀

Leonov (left) with Deke Slayton in the Soyuz spacecraft

Interview with Neil Armstrong

First Human to Walk on the Moon

Introduction

On July 20, 1969, the world held its breath. Hundreds of millions of people gathered around televisions and radios across continents, waiting for a moment that had never before occurred in human history. High above the Moon's silent landscape, a fragile spacecraft descended toward the surface carrying two astronauts and the hopes of an entire planet.

At the controls of the lunar module ***Eagle*** *was a quiet, disciplined engineer and test pilot from Ohio: Neil Armstrong. The mission had already pushed the limits of technology and human endurance. Just moments before landing, alarms sounded inside the spacecraft's computer. Fuel was running dangerously low. The designated landing site was covered with boulders. Armstrong took manual control of the lunar module and guided it across the Moon's surface, searching for a safe place to land. Seconds later, the words that would echo through history were transmitted to Earth: "Houston, Tranquility Base here. The Eagle has landed." Humanity had arrived on another world.*

Hours later, Armstrong slowly descended the ladder of the lunar module. Beneath him stretched a gray, untouched landscape that had remained unchanged for billions of years. As he stepped onto the surface of the Moon, he spoke words that would become one of the most famous phrases ever recorded: "That's one small step for man, one giant leap for mankind."

In that moment, the boundary between imagination and reality disappeared. For the first time, a human being was standing on a celestial body beyond Earth. Yet Armstrong himself never sought fame. Known for his humility and quiet professionalism, he often reminded people that the Moon landing was not the achievement of one individual but the result of the efforts of thousands of scientists, engineers, and technicians who made the Apollo program possible.

Before becoming the first person to walk on the Moon, Armstrong had already built an extraordinary career. As a naval aviator, he flew combat missions during the Korean War. Later, as a test pilot, he pushed experimental aircraft to their limits, helping expand the frontiers of aviation. His calm judgment and technical skill made him one of NASA's most trusted astronauts.

But it was the Apollo 11 mission that forever placed his name among the greatest explorers in history. Neil Armstrong represented a generation of pioneers who believed that humanity's destiny extended beyond Earth.

Their courage transformed space exploration from a distant dream into a reality.

More than half a century later, his footsteps remain on the Moon—silent imprints in lunar dust, untouched by wind or weather. They serve as a lasting reminder that human curiosity and determination can carry us far beyond the limits once thought impossible. In the following conversation, we revisit the reflections of the man who took humanity's first step onto another world.

Neil Alden Armstrong (August 5, 1930 – August 25, 2012) was an American astronaut and aeronautical engineer who, as the commander of the 1969 Apollo 11 mission, became the first person to walk on the Moon. He was also a naval aviator, test pilot and university professor.

1. Commander Armstrong, when did you first begin dreaming about flying and exploring beyond Earth?

From a very early age I was fascinated by flight. My father took me to air shows when I was a boy in Ohio, and I remember watching airplanes climb into the sky and wondering what it would feel like to be up there. By the time I was sixteen, I had already begun taking flying lessons. At that point, aviation was my world. Space exploration came later, but it felt like a natural extension of the same dream—to explore the unknown skies beyond our planet.

---

2. What did you feel in the moments before the launch of Apollo 11?

The launch of a Saturn V rocket is an extraordinary event. The vehicle weighs thousands of tons, and when the engines ignite, the entire structure trembles with tremendous power. But during those moments, our focus was primarily on procedures and checklists. Years of training prepared us for that moment. While the experience was certainly thrilling, our attention was fixed on doing the job correctly.

3. When you first looked at Earth from space, what went through your mind?

When you see Earth from space, the first feeling is not triumph — it is perspective. The planet appears small, delicate, almost fragile against the vast darkness of space. From the Moon, Earth is not a map of nations or borders. It is a single blue and white sphere suspended in silence.

What struck me most was its beauty. The oceans shimmered in deep blue, clouds swirled like brushstrokes across the surface, and the continents seemed peaceful and unified. All the conflicts, politics, and divisions that occupy our daily lives were invisible from that distance.

It reminded me that humanity shares one home. Every person who has ever lived, loved, struggled, or dreamed exists on that small world. Seeing it from afar makes you realize how important it is to protect it. In that moment, I did not feel that we had conquered space. Instead, I felt that space had given us a gift — the ability to see our own planet with new humility and appreciation.

And perhaps that is one of the greatest legacies of space exploration: it helps us understand that while we may explore distant worlds, Earth will always remain our most precious one.

---

4. Commander Armstrong, when you stepped onto the lunar surface during the Apollo 11 mission, it became one of the most famous moments in human history. What was going through your mind at that instant?

At that moment, my thoughts were primarily focused on the task at hand. Landing on the Moon requires an enormous amount of preparation and teamwork. Thousands of engineers, scientists, and technicians had worked for years to make the mission possible. As I climbed down the ladder of the lunar module **Apollo Lunar Module Eagle**, I was very aware that the success of the mission depended on following our procedures carefully.

When I placed my foot on the lunar surface, I was thinking about how remarkable it was that the spacecraft had landed successfully and that we were finally able to carry out the exploration we had trained for so long. The surface itself was slightly different from what we had expected. The lunar soil was fine and powdery, almost like very soft dust. I wanted to make sure that the footing was stable before moving farther away from the ladder.

Of course, I also realized that the moment represented something much larger than our mission. It was not simply the achievement of one astronaut or even one nation, it was the result of an extraordinary effort by many people working together to expand human knowledge.

That is why I said that it was **"one small step for a man, one giant leap for mankind."** The words were meant to reflect the idea that although the step itself was small, the achievement represented a major milestone in the history of human exploration. Looking back, what I remember most clearly is the sense that we were witnessing the beginning of a new chapter in humanity's journey beyond Earth.

The Apollo 11 crew: Armstrong, Michael Collins, and Buzz Aldrin.

5. The Apollo missions involved enormous risks. How did you approach the dangers of spaceflight?

Spaceflight in those days was inherently dangerous, and every astronaut understood that from the beginning. The rockets, the spacecraft, and even the procedures were still being developed. We were pushing the boundaries of engineering and human experience at the same time.

But we did not approach the mission recklessly. On the contrary, every aspect of the program was built on discipline, preparation, and trust. Thousands of engineers, scientists, and technicians worked tirelessly to reduce risks and test every system. By the time we climbed into the spacecraft, we knew that an extraordinary team stood behind us.

Astronauts are trained to think like test pilots. You study the systems, you understand the procedures, and you prepare for contingencies. Fear is not something you ignore, but something you manage through knowledge and training. When the countdown begins, your focus shifts entirely to the job at hand.

There were moments during the Apollo missions when decisions had to be made quickly and calmly. The key was always to remain disciplined and rely on both your training and your team.

Ultimately, we accepted the risks because we believed in the purpose of the mission. Exploring space was not only about reaching the Moon; it was about expanding human knowledge and demonstrating what humanity could achieve when it works together.In that sense, every astronaut understood that we were part of something much larger than ourselves.

Armstrong and X-15-1 after a research flight in 1960

---

6. What surprised you most about the Moon?

What surprised me most about the Moon was its profound stillness. When you stand on the lunar surface, there is a sense of complete silence unlike anything on Earth. There is no atmosphere, no wind, no sound — only the quiet hum of your life-support system inside the suit.

Another surprise was the texture of the surface. From orbit, the Moon appears smooth and almost gentle. But when you walk on it, the ground is covered with fine, powdery dust called lunar regolith. It clings to everything — boots, suits, and equipment — almost like very soft graphite powder.

The horizon also looked different. Because the Moon is much smaller than Earth, the horizon seems closer and more sharply curved. It gives the landscape a strange, almost theatrical appearance, as if you are standing on a small stage surrounded by darkness.

Recovery of Gemini 8 from the western Pacific Ocean; Armstrong sitting to the right

But perhaps the most remarkable sight was the sky. Even during the lunar day, the sky above is completely black. Against that darkness, Earth appears bright and beautiful — a glowing blue sphere hanging in space.

Standing there, you realize that the Moon itself is quiet and lifeless, but it offers one of the most extraordinary viewpoints in the universe: the chance to look back at our own world and see it as a single, fragile home for all humanity.

---

7. You spent only a short time on the Moon, yet that moment changed history. Did you realize the impact the mission would have?

At the time, our focus was entirely on the mission itself. There were many technical tasks to complete, experiments to deploy, and procedures to follow. When you are in that environment, you concentrate on doing your job correctly and safely.

Of course, we understood that the mission carried historical significance. The effort that led to the Moon landing involved hundreds of thousands of people—engineers, scientists, technicians, and support teams across the country. We knew we were representing their work and the aspirations of an entire generation.

But the full global impact was something that became clearer only afterward. When we returned to Earth, we began to hear how people around the world had watched the landing. For a moment, humanity seemed united in curiosity and wonder.

I have always believed that the landing was less about the astronauts who walked on the Moon and more about what people on Earth achieved together. It demonstrated that when nations invest in science, education, and exploration, remarkable things can happen.

If the mission inspired young people to pursue science, engineering, and discovery, then perhaps its greatest legacy lies not in the footprints on the Moon, but in the minds, it encouraged to dream beyond our planet.

Neil Armstrong photographed by Buzz Aldrin after the completion of the Lunar EVA on the Apollo 11 flight.

---

8. Commander Armstrong, some conspiracy theories claim that film director Stanley Kubrick secretly helped NASA stage the Apollo 11 Moon landing in a studio, supposedly in case the real mission failed. Could you clarify whether there is any truth to this claim?

No, that claim is not true. The Apollo 11 mission was a real spaceflight that successfully landed astronauts on the Moon on July 20, 1969. The idea that the landing was staged by a film director such as Stanley Kubrick is a popular conspiracy theory, but it has no factual basis.

The mission was tracked in real time by numerous independent organizations around the world, including observatories and space agencies in countries outside the United States. During the Cold War, the Soviet Union carefully monitored American space missions. If the Moon landing had been fabricated, Soviet scientists and engineers would almost certainly have exposed it immediately.

In addition, the Apollo missions returned hundreds of kilograms of lunar rocks and soil that have been studied by scientists across many nations. These samples have properties that clearly distinguish them from any material found on Earth.

Astronauts also placed scientific instruments on the lunar surface, including laser reflectors that are still used today to measure the precise distance between Earth and the Moon.

More recently, spacecraft orbiting the Moon—such as NASA's Lunar Reconnaissance Orbiter—have photographed the Apollo landing sites, showing the descent stages of the lunar modules and even the paths astronauts walked on the surface.

The suggestion that a filmmaker secretly staged the landing misunderstands both the scale and the transparency of the Apollo program. The effort involved hundreds of thousands of engineers, technicians, and scientists working in many different locations. It would have been impossible to conceal a fabrication of that magnitude.

Stanley Kubrick was an extraordinary filmmaker, but he had nothing to do with the Apollo missions. The Moon landing was a real achievement made possible by the dedication and hard work of an enormous team of people who helped humanity reach another world.

Question to director Stanly Kubrick:

Mr. Kubrick, conspiracy theories have circulated for decades claiming that you secretly helped the U.S. government and NASA stage the Moon landing in a film studio for the Apollo 11 Moon Landing. Did you ever have any involvement in such an effort, and how do you respond to these claims?

Stanley Kubrick: I have heard those theories many times, and I must admit they say more about human imagination than about filmmaking or space exploration.

During the 1960s I was indeed working on the film *2001: A Space Odyssey*, which aimed to depict space travel with a level of realism that audiences had never seen before. My team worked very closely with scientists, engineers, and aerospace consultants to create visuals that felt authentic. Perhaps that realism led some people to imagine that cinema could substitute for reality.

But filmmaking and space exploration are very different enterprises. Creating convincing imagery on a screen is one thing; launching a spacecraft, sending astronauts hundreds of thousands of miles into space, landing on another world, and returning them safely is quite another.

The Apollo missions involved hundreds of thousands of engineers, technicians, scientists, and astronauts. The idea that such an enormous achievement could be secretly staged inside a film studio misunderstands both the complexity of the mission and the nature of filmmaking.

I always believed that the real universe is far more extraordinary than anything we could construct on a set. My goal as a filmmaker was to help audiences imagine humanity's future in space—not to replace the remarkable accomplishments that scientists and explorers were already achieving. If anything, the Moon landing proved that reality could surpass even the most ambitious science fiction.

9. **Commander Armstrong, over the years some rumors and conspiracy theories have claimed that the Apollo 11 Moon landing never actually happened. Some individuals even claim to have "proven" that the mission was staged. Could you share your perspective on these claims and explain what really occurred during the Apollo 11 mission?**

I'm aware that such claims exist, and from time to time they attract public attention. However, the reality of the Apollo 11 mission is supported by an overwhelming amount of scientific, technical, and historical evidence.

The Apollo program involved more than **400,000 engineers, scientists, technicians, and support personnel** working across thousands of organizations. Designing the Saturn V rocket, developing the lunar module, building navigation systems, and coordinating mission operations required the efforts of an enormous international scientific community. It would be impossible to fabricate an undertaking of that scale without it becoming immediately obvious.

In addition, the evidence for the Moon landing is extensive. We brought back **382 kilograms of lunar rock and soil samples** from the Apollo missions, which have been studied by scientists around the world for decades. These materials have chemical and geological properties that differ significantly from any rocks found on Earth.

During the missions, we also placed scientific instruments on the lunar surface, including laser reflectors that are still used today by scientists to measure the distance between the Earth and the Moon with great precision. Observatories around the world have used these reflectors continuously since the Apollo missions.

The flights were tracked not only by NASA but also by independent observatories and space agencies in other countries, including those that were not allies of the United States during the Cold War. If the missions had been fabricated, those organizations would certainly have exposed the deception.

The photographs, video transmissions, telemetry data, and physical artifacts of the Apollo missions are all consistent with a real lunar landing. More recently, spacecraft orbiting the Moon have even photographed the Apollo landing sites, including the descent stages of the lunar modules and the tracks left by astronauts.

In science and engineering, extraordinary claims require extraordinary evidence. The claim that the Moon landing was faked does not withstand serious examination.

The Apollo missions were not the achievement of one astronaut or even a small group of astronauts. They were the result of the dedication and ingenuity of hundreds of thousands of people who worked tirelessly to make the exploration of the Moon possible.

I was fortunate to represent that effort by taking the first step onto the lunar surface, but the achievement truly belonged to all of them—and to humanity.

---

10. Many astronauts say that seeing Earth from space changes how they think about humanity. Did it change your perspective?

Yes, I believe it does change your perspective in a profound way. When you see Earth from space, it appears as a small, brilliant sphere suspended in the darkness. From that distance, you cannot see borders, conflicts, or divisions between nations. You see only one planet — our shared home.

That view reminds you how interconnected humanity really is. Every civilization, every culture, every person who has ever lived exists on that single world. Realization brings a sense of humility. Our planet is beautiful, but it is also fragile.

During the mission of Apollo 11 Moon Landing, there were moments when we could look back and see Earth rising above the lunar horizon. It was a striking image — a bright blue and white oasis in the vast blackness of space.

Seeing that view reinforces the idea that exploration should unite us rather than divide us. Space exploration represents one of the best examples of what humanity can achieve when we pursue knowledge and discovery together. In that sense, traveling into space does not make you feel distant from Earth. It makes you feel more connected to it — and more aware of how important it is to protect the only home we have.

The Apollo 11 crew and President Richard Nixon during the post-mission quarantine period

11. Humanity is now preparing to return to the Moon and eventually travel to Mars. What are your hopes for the future of space exploration?

I believe exploration is a fundamental part of human nature. Each generation builds upon the achievements of the previous one. Returning to the Moon and traveling to Mars are logical next steps. These missions will expand scientific knowledge and inspire future explorers.

---

12. During your career you met many presidents, world leaders, and influential figures. Which individuals impressed you the most, and why?

Over the years I had the opportunity to meet many remarkable people—presidents, scientists, engineers, and leaders from around the world. What impressed me most was not necessarily their power, but their vision.

One person who left a lasting impression was **John F. Kennedy**. His decision to commit the nation to landing a man on the Moon within the decade was an extraordinary example of leadership. At the time, the goal seemed almost impossible, yet his vision inspired a generation of scientists, engineers, and astronauts to work toward something that had never been done before.

I was also impressed by the dedication of many engineers and researchers within NASA. Although they were not always in the public spotlight, their brilliance and determination made the Apollo missions possible. The success of the Moon landing was truly the achievement of hundreds of thousands of people.

Ticker tape parade for the Apollo 11 astronauts. Location is Manhattan, New York City on the section of Broadway known as the "Canyon of Heroes". Pictured in the lead car, from the left, waving their hands, are astronauts Edwin E. Aldrin Jr., Michael Collins and Neil A. Armstrong

Later, during our worldwide goodwill tour after the Apollo 11 Moon Landing, we met leaders and citizens from many different countries. What struck me most was how people everywhere felt a sense of pride in the achievement. It reminded me that exploration transcends politics and national boundaries.

In truth, the individuals who impressed me most were those who combined intelligence with humility—people who understood that great accomplishments are rarely the work of one person alone, but of many working together toward a common goal.

President Barack Obama poses with the Apollo 11 crew on the 40th anniversary of the Apollo 11 lunar landing, July 20, 2009: Buzz Aldrin, Michael Collins, and Neil Armstrong.

---

13. If you could share one message with future generations looking toward the stars, what would it be?

My message would be simple: never stop exploring. Human progress has always depended on curiosity, the desire to ask questions about the unknown and to venture beyond the boundaries of what we already understand. Space exploration is simply the latest chapter in that long story of discovery.

When we traveled to the Moon during the Apollo 11 Moon Landing, we were not just exploring another world. We were demonstrating what humanity can accomplish when knowledge, courage, and cooperation come together.

Future generations will go farther than we ever did. They will travel to Mars, explore distant planets, and perhaps one day journey beyond our solar system. But as they look outward toward the stars, they should also remember to look back at Earth. It remains our home, and its well-being is inseparable from the future of exploration.

Exploration is not only about reaching new places. It is about expanding human understanding and reminding us that our greatest achievements come when we work together for a common purpose. The stars will always be there, waiting. The question is whether we will continue to have the curiosity and determination to reach them.

Interview with Michael Collins

Command Module Pilot of Apollo 11

Introduction

On July 20, 1969, the world watched as two astronauts descended toward the surface of the Moon. History would remember the moment when Neil Armstrong stepped onto the lunar surface and spoke words that would echo across generations. But high above the Moon, another astronaut was performing a mission just as critical—and perhaps even more solitary.

Michael Collins remained in lunar orbit aboard the ***Apollo 11 command module Columbia****, circling the Moon while his crewmates explored its surface. As the spacecraft passed behind the Moon during each orbit, Collins was completely cut off from radio communication with Earth. For nearly half an hour at a time, he was the most isolated human being in existence.*

At those moments, Collins was truly alone—separated not only from Earth but also from the other two members of his crew. Beneath him stretched the silent lunar landscape, while above him hung the distant blue planet that was home to all humanity. Yet Collins understood the importance of his role.

Without the command module pilot, the mission could never have succeeded. While Armstrong and Aldrin explored the Moon, Collins carefully monitored the spacecraft's systems and prepared for the critical task of reuniting with the lunar module in orbit. If anything had gone wrong, he would have been responsible for bringing his crewmates safely back to Earth.

Michael Collins had already built an impressive career as a pilot and astronaut before Apollo 11. A graduate of the U.S. Air Force Test Pilot School, he flew combat missions and later joined NASA during the Gemini program. On ***Gemini 10****, he performed two spacewalks and gained valuable experience operating spacecraft in orbit. But it was Apollo 11 that secured his place in history.*

While Armstrong and Aldrin left the first human footprints on the Moon, Collins remained the silent guardian of the mission—ensuring that humanity's great journey would end safely. Though he often described himself modestly as the "forgotten astronaut" of Apollo 11, many historians recognize that the success of the mission depended just as much on his skill and calm professionalism. In the vast silence of lunar orbit, Michael Collins played one of the most essential roles in humanity's first journey to another world.

Michael Collins (October 31, 1930 – April 28, 2021) was an American astronaut who flew the Apollo 11 command module *Columbia* around the Moon in 1969 while his crewmates, Neil Armstrong and Buzz Aldrin, made the first crewed landing on the surface. He was also a test pilot and major general in the U.S. Air Force Reserve.

---

1. Colonel Collins, when did your interest in aviation begin?

My interest in aviation developed gradually. I was inspired by the early achievements of military aviation and decided to pursue a career as a pilot. Eventually I became a test pilot, which naturally led to the opportunity to join NASA's astronaut program.

2. What did it mean to be selected as part of the Apollo program?

Being selected for the Apollo program was both an extraordinary honor and a tremendous responsibility. At the time, space exploration was advancing at a remarkable pace, and the goal of landing humans on the Moon represented one of the most ambitious undertakings in history.

For an astronaut, selection meant becoming part of something far larger than oneself. Thousands of scientists, engineers, and technicians were working across the country to make the missions possible. Our role as astronauts was simply the visible tip of a very large and dedicated team.

The training was intense and demanding. We studied spacecraft systems, navigation, geology, orbital mechanics, and emergency procedures. Every detail mattered because the mission would take us farther from Earth than any humans had ever traveled before.

Personally, I felt a deep sense of excitement. The Apollo program represented the frontier of exploration. To be part of the mission that would attempt to land humans on the Moon during the Apollo 11 Moon Landing was an opportunity that very few people in history could ever experience.

At the same time, there was a strong sense of humility. None of us could accomplish such a mission alone. Being selected meant becoming part of a team that trusted one another completely, knowing that success—and sometimes survival—depended on that trust.

---

3. During Apollo 11, you remained in lunar orbit while Armstrong and Aldrin descended to the Moon. What was that experience like?

My role during the mission of Apollo 11 Moon Landing was to pilot the command module *Columbia* while Neil Armstrong and Buzz Aldrin descended to the lunar surface in the lunar module *Eagle*. It was a unique position—one that placed me both close to the historic landing and yet very much alone.

As I orbited the Moon, there were moments when the spacecraft passed behind the far side of the Moon. During those periods, I was completely out of radio contact with both Earth and my crewmates. For roughly forty-five minutes each orbit, I was quite literally the most isolated human being in existence.

Yet I never felt lonely. My attention was focused entirely on the spacecraft systems and the responsibilities of operating the command module. I knew that when Neil and Buzz returned from the surface, it would be my job to rendezvous with them in orbit. Everything had to work perfectly.

There was also a sense of quiet awe. As I looked out the window, I could see the stark beauty of the lunar surface passing slowly beneath me, and beyond it the distant Earth shining in space. It was a powerful reminder of how small and precious our home planet is.

I often say that Apollo 11 was not just about the two men who walked on the Moon. It was about the entire team and the extraordinary journey that humanity took together during that mission.

Collins (right) with John Young (left) and a model of their Gemini spacecraft and Titan II booster

---

4. Some people have called you "the loneliest man in history." Did you feel lonely during the mission?

That description has followed me for many years, but it was never quite accurate. During the mission of Apollo 11 Moon Landing, my job was to pilot the command module

Columbia while Neil Armstrong and Buzz Aldrin explored the lunar surface.

It is true that when I passed behind the far side of the Moon, I was temporarily cut off from radio communication with both Earth and my crewmates. For those periods, I was the only human not in contact with anyone else. But loneliness is a feeling, and I did not feel lonely.

I was busy. There were systems to monitor, navigation calculations to check, and preparations to make for the rendezvous with the lunar module when Neil and Buzz lifted off from the Moon. My focus was entirely on the mission.

In fact, I felt a deep sense of contentment and responsibility. I knew that my role was essential. If anything had gone wrong with the command module, my crewmates would not have been able to return home.

As I looked out the window and saw the Moon beneath me and the Earth shining in the distance, I felt privileged to witness something few humans ever had. Rather than loneliness, the feeling was one of quiet awe and profound appreciation for the remarkable journey we were undertaking together.

---

5. What did Earth look like from lunar orbit?

From the distance of the Moon, Earth appears as a small, brilliant blue sphere suspended in the darkness of space. It is beautiful and delicate, and it reminds you how small our planet is in the vast universe.

Agena Target Docking Vehicle photographed near the Gemini 10 spacecraft, photographed by NASA – Public Domain

---

6. Were you ever concerned about the possibility that the lunar module might not return from the surface?

Of course, that possibility existed. Spaceflight is a complex and demanding endeavor, and everyone involved in the Apollo program understood that there were risks at every stage of the mission.

When Neil Armstrong and Buzz Aldrin descended to the lunar surface during the Apollo 11 Moon Landing, I remained in lunar orbit aboard the command module *Columbia*. My responsibility was to maintain the spacecraft and prepare for the rendezvous when the lunar module lifted off from the Moon.

We trained extensively for every possible contingency. Engineers and mission planners have carefully designed systems and procedures to maximize safety and reliability. Still, we were aware that exploration always involves uncertainty.

Collins (center) with William Anders (left) and Frank Borman (right)

Personally, I tried not to dwell on worst-case scenarios. My focus was on doing my job correctly and being ready to bring my crewmates home. If the lunar module had failed to return—which fortunately it did not—it would have been a tragic outcome. But that possibility was something we all accepted as part of the mission.

What mattered most was trust: trust in the spacecraft, trust in the engineers and scientists who designed it, and trust in the abilities of my fellow astronauts. That trust is what allowed us to undertake such an extraordinary journey with confidence and determination.

---

7. What moment during the Apollo 11 mission stands out most in your memory?

There were many extraordinary moments during the mission of Apollo 11 Moon Landing, but the one that stands out most vividly for me was the moment when Neil Armstrong and Buzz Aldrin lifted off from the lunar surface to rejoin me in lunar orbit.

While they were on the Moon, my responsibility was to pilot the command module *Columbia* and maintain a precise orbit around the Moon. Everything depended on the successful launch of the lunar module and our ability to rendezvous in space.

When I first saw the ascent stage of the lunar module rising from the Moon and approaching the command module, it was an incredibly satisfying moment. It meant that the most difficult part of the mission had succeeded and that we would soon be reunited as a crew.

That moment also carried a deep sense of relief. Space missions involve countless complex operations, and every step must work exactly as planned. Seeing the lunar module safely return to orbit confirmed that our journey could continue toward home.

Looking back, the Apollo mission was filled with historic milestones, but for me the most memorable moment was that quiet reunion in orbit—three astronauts together again after humanity's first visit to another world.

---

8. How did the Apollo 11 mission change your perspective on humanity?

Seeing Earth from such a great distance changes the way you think about our planet. It appears peaceful and unified, without visible borders. It reminds you that humanity shares one small world.

---

9. What do you believe was the greatest achievement of the Apollo program?

Many people would say that the greatest achievement of the Apollo program was landing humans on the Moon during the Apollo 11 Moon Landing. And certainly, that moment captured the imagination of the world.

But in my view, the real achievement was something even larger. The Apollo program demonstrated what humanity can accomplish when knowledge, determination, and cooperation come together. Hundreds of thousands of engineers, scientists, technicians, mathematicians, and astronauts—worked together toward a single goal that once seemed impossible.

The technological advances were remarkable: powerful rockets, sophisticated spacecraft, guidance computers, and entirely new methods of navigation and communication. These innovations did not only serve space exploration; they also influenced science, technology, and engineering here on Earth.

Equally important was the inspiration the program provided. For millions of people around the world, Apollo showed that exploration is still part of the human spirit. It encouraged young people to pursue careers in science and engineering and reminded society that bold goals can unite us.

In that sense, the greatest achievement of Apollo was not only reaching the Moon. It proved that humanity, when it works together, can reach far beyond what once seemed possible.

---

10. What message would you share with future explorers who will travel to the Moon and Mars?

To the explorers who will journey to the Moon again and one day travel onward to Mars, I would say this: remember that exploration is one of humanity's most noble endeavors.

During the mission of Apollo 11 Moon Landing, we were aware that we were opening a door rather than completing a journey. The Moon landing was only the beginning of humanity's venture beyond Earth.

Collins in the command module simulator

Future explorers will face challenges far greater than those we encountered. Traveling to Mars, for example, will require longer missions, new technologies, and tremendous endurance. But the spirit that carried humans to the Moon—the curiosity to explore the unknown—will guide those journeys as well.

As you travel to distant worlds, never forget the small blue planet from which you came. Earth is our home, and every voyage into space reminds us how precious and fragile it is. Explore boldly, learn humbly, and work together. The universe is vast, and the future of discovery belongs to those who have the courage to pursue it.

Interview with Pete Conrad

Commander of Apollo 12

Introduction

Only four months after humanity first set foot on the Moon, NASA prepared for a second lunar landing. The mission was ***Apollo 12****, and its commander was an astronaut known not only for his exceptional flying skills but also for his humor and fearless personality—Pete Conrad.*

Unlike the tense and historic first landing of Apollo 11, Apollo 12 aimed to demonstrate precision and confidence. The goal was to land in a very specific location on the Moon, near the robotic spacecraft ***Surveyor 3****, which had landed there two years earlier. But the mission began with a dramatic test of nerve.*

Just seconds after liftoff, lightning struck the Saturn V rocket—twice. The spacecraft's electrical systems temporarily failed, and warning lights flashed throughout the cockpit. For a moment, it appeared the mission might end before it had truly begun. Calmly and methodically, Conrad and his crew followed procedures that restored the spacecraft's systems. Within minutes, Apollo 12 was safely on its way to the Moon.

Days later, Conrad guided the lunar module ***Intrepid*** *to one of the most precise landings in the history of lunar exploration. When he stepped onto the Moon on November 19, 1969, he became the third human being to walk in another world. True to his personality, Conrad's first words on the lunar surface were playful:*

"Whoopee! Man, that may have been a small one for Neil, but that's a long one for me." His remark humorously acknowledged that he was shorter than Neil Armstrong, while also celebrating the incredible achievement of returning to the Moon so soon after the first landing.

During their mission, Conrad and astronaut Alan Bean conducted scientific experiments, collected lunar samples, and even examined parts of the Surveyor spacecraft—demonstrating how human exploration could work alongside robotic missions. Pete Conrad represented a new phase of lunar exploration: confident, precise, and increasingly ambitious.

His skill as a pilot and his ability to remain calm under pressure helped ensure that the second lunar landing was a complete success. While the world often remembers the first steps on the Moon, the achievements of Apollo 12 proved that humanity could return—and again—to explore other worlds.

Charles "Pete" Conrad Jr. (June 2, 1930 – July 8, 1999) was an American NASA astronaut, aeronautical engineer, naval officer, aviator, and test pilot who commanded the Apollo 12 mission, on which he became the third person to walk on the Moon. Conrad was selected for NASA's second astronaut class in 1962.

---

1. Commander Conrad, when did your passion for aviation begin?

I became interested in flying at a young age. Aviation offered a sense of freedom and challenge that fascinated me. Eventually I attended the U.S. Naval Academy and became a naval aviator, which led to my work as a test pilot.

2. How did you feel when NASA selected you as one of the early astronauts?

It was an incredible opportunity. The early space programs were pushing the limits of technology, and being part of that effort meant helping to define the future of human exploration.

---

3. Your Apollo 12 mission experienced lightning strikes shortly after launch. What was going through your mind at that moment?

The launch of Apollo 12 Moon Landing was already intense, but just seconds after liftoff something very unusual happened. As the Saturn V rocket climbed through the clouds, lightning struck the vehicle—twice.

Inside the spacecraft, we suddenly saw many warning lights and alarms appear on the control panels. For a moment it looked as if several of the spacecraft's electrical systems had failed. The telemetry being sent to Mission Control became erratic, and even some of our instruments stopped working properly.

My first thought was to stay calm and figure out what was actually happening. As astronauts, we were trained to focus on procedures rather than panic. We quickly started checking the systems and waiting for guidance from Mission Control.

One of the most critical moments came when a flight controller on the ground suggested a solution that became famous: "SCE to AUX." That simple switch restored the spacecraft's electrical systems and allowed us to regain normal telemetry.

Once things stabilized, we realized the rocket itself was still performing normally and the mission could continue. It was a reminder that spaceflight is never routine and that even small technical decisions—like knowing the right switch to flip—can make the difference between aborting a mission and continuing all the way to the Moon.

In the end, the experience reinforced something we all understood about exploration: unexpected challenges will always appear, but preparation, teamwork, and calm decision-making are what allow you to overcome them. 🚀

Conrad preparing for water egress training in the Gemini Static Article 5 spacecraft

4. What was it like landing on the Moon compared with earlier space missions?

Landing on the Moon during the Apollo 12 Moon Landing was very different from my earlier experiences in space. In previous missions, such as the Gemini 11 mission, we spent our time orbiting the Earth, conducting rendezvous maneuvers, performing experiments, and learning how humans and spacecraft could operate in orbit.

The Moon landing, however, was something entirely new. Instead of simply circling a planet, we were guiding a spacecraft down to the surface of another world. As the lunar module *Intrepid* descended, every moment required careful attention—monitoring altitude, velocity, and the landing site.

The most striking difference was the sense of arrival. In orbit, Earth is always below you, moving silently beneath the spacecraft. But on the Moon, when the engines shut down and the dust settles, you realize you are standing somewhere no human being has ever stood before.

The environment was also extraordinary. The sky was completely black, even in daylight, and the landscape looked almost untouched—craters, rocks, and fine gray dust stretching to the horizon. But what made the experience truly remarkable was the realization that humanity had extended its reach beyond Earth. Walking on the Moon was not just another mission was the opening of a new frontier for exploration.

5. Apollo 12 achieved an extremely precise landing near the Surveyor spacecraft. Why was that important?

One of the key goals of the mission of Apollo 12 Moon Landing was to demonstrate that we could land the lunar module with great precision. Unlike the first Moon landing, which focused primarily on proving that a landing was possible, Apollo 12 aimed to show that we could reach a specific target.

Our target was the robotic spacecraft Surveyor 3, which had landed on the Moon two years earlier in the Ocean of Storms. By landing only a short distance away, we could examine the spacecraft and bring some of its parts back to Earth.

This was scientifically valuable for several reasons. First, engineers wanted to study how materials survived after being exposed to the harsh lunar environment for more than two years—extreme temperatures, radiation, and micrometeorites.

Second, the precision landing demonstrated that astronauts could safely guide spacecraft to predetermined locations on the Moon. This capability was essential for future missions that would need to reach scientifically interesting areas rather than simply landing wherever conditions were safest.

When we walked over to Surveyor 3 and examined it, it felt almost like visiting an old robotic colleague that had been waiting there for us. In many ways, that precise landing proved that lunar exploration had progressed from simply reaching the Moon to beginning serious scientific exploration of its surface.

Conrad (right) with his Gemini 11 crewmate Dick Gordon, following their flight

---

6. What was the most memorable part of your time on the Moon?

There were many unforgettable moments during the mission of Apollo 12 Moon Landing, but what stands out most in my memory is the feeling of freedom and excitement as we began exploring the lunar surface.

After the intense concentration required during the descent and landing of the lunar module *Intrepid*, stepping outside onto the Moon felt almost surreal. The landscape stretched out in every direction—gray, silent, and untouched. You realized very quickly that you were standing on a world where no human had ever walked before.

One particularly memorable experience was walking over to the robotic spacecraft Surveyor 3. It had landed there two years earlier and seeing it sitting quietly on the lunar surface felt like discovering a small piece of human history waiting for us.

But beyond the scientific work and the historic significance, there was also a sense of joy. The Moon's gravity is only about one-sixth that of Earth, which meant that moving and walking felt almost like slow-motion hopping. It was an extraordinary sensation—both playful and humble at the same time.

Looking back, the most memorable part was simply the realization that humanity had extended its reach to another world. For a moment, we were not just visitors from Earth—we were explorers walking across the surface of the Moon. 🚀

---

7. How did seeing Earth from the Moon affect your perspective?

From the Moon, Earth appears small and distant luminous blue sphere suspended in the vast darkness of space. In that moment, you realize how interconnected humanity truly is.

From such a perspective, the divisions that seem so important on Earth—borders, nations, and conflicts—simply disappear.

Our planet is revealed for what it really is: a single world shared by all people. Every civilization, every culture, every person who has ever lived exists on that fragile sphere. Seen from the Moon, Earth looks peaceful, beautiful, and remarkably delicate.

That view leaves a lasting impression on astronauts. It reminds us that despite our differences, humanity shares one home and one future. Protecting that home—and learning to work together—may be the most important lesson that space exploration can teach us. 🌍🚀

---

8. What do you think was the greatest achievement of the Apollo missions?

The Apollo missions proved that humans could travel to another celestial body and return safely. That achievement required extraordinary teamwork among scientists, engineers, and astronauts.

Conrad, Dick Gordon, and Alan Bean pose with their Apollo 12 Saturn V Moon rocket in the background.

This is the official crew insignia for Apollo 12, the United States' second lunar landing mission.

9. Commander Conrad, it is often said that your famous first words on the Moon were connected to a bet you made with Italian journalist Oriana Fallaci. Could you tell us more about that story?

Yes, that story is true. Before the mission of Apollo 12 Moon Landing, I had a conversation with the Italian journalist Oriana Fallaci. At the time, there was a popular conspiracy theory suggesting that NASA had scripted Neil Armstrong's famous first words on the Moon.

Fallaci wondered whether astronauts were told exactly what to say when they stepped onto the lunar surface. I told her that wasn't the case at all. NASA certainly didn't script our first words.

So, I jokingly proposed a bet. I told her that when I stepped onto the Moon, I would say something completely different from Armstrong's famous line and prove that astronauts were free to speak naturally. When the moment came and I climbed down the ladder of the lunar module *Intrepid*, I looked down at the lunar surface and said:

"Whoopee! Man, that may have been a small one for Neil, but that's a long one for me."

The reason for the joke was simple—Neil Armstrong was quite tall, while I was considerably shorter. My step down to the surface was physically longer than his had been.

The remark wasn't scripted. It was simply a lighthearted moment during a historic mission. Space exploration can be very serious work, but sometimes it's good to remind people that astronauts are human too—and that even on the Moon, a little humor can travel a long way.

---

10. What message would you share with future astronauts exploring the Moon and Mars?

To the astronauts who will one day return to the Moon and travel onward to Mars, I would say this: you are continuing a journey that began long before any of us were born.

When we walked on the Moon during the mission of Apollo 12 Moon Landing, we understood that we were only taking the first steps into a much larger future of exploration. The Moon landings were not the end of the story—they were the opening chapter.

Future explorers will face challenges that are even greater than ours. Traveling to Mars will require longer missions, new technologies, and extraordinary patience and courage. But the spirit that drives exploration—the desire to discover what lies beyond the horizon—remains the same.

As you explore distant worlds, remember where you came from. Earth will always be humanity's home, a small and beautiful planet in a vast universe. Every journey into space reminds us how precious that home really is.

Explore with curiosity, humility, and determination. The footprints we left on the Moon were only temporary marks in the dust. The true legacy of exploration will belong to those who continue the journey and carry humanity farther into the universe. 🚀

Conrad descends the Lunar Module ladder, moments before becoming the third human to walk on the Moon
Photographed by Alan Bean, NASA – Public Domian

Interview with Edgar Mitchell

Lunar Module Pilot of Apollo 14

Introduction

By 1971, humanity had already achieved what once seemed impossible. Astronauts had traveled to the Moon, walked upon its surface, and returned safely to Earth. Yet each mission still carried enormous challenges and risks. The mission of ***Apollo 14*** *was particularly important.*

Just one year earlier, the world had watched the dramatic rescue of the crew of Apollo 13 after an explosion crippled their spacecraft on the way to the Moon. That crisis forced NASA to suspend lunar missions temporarily while engineers investigated what had gone wrong. Apollo 14 would demonstrate that the path to the Moon remained open.

Commanded by Alan Shepard, the mission included lunar module pilot Edgar Mitchell—an astronaut known for both his scientific discipline and his thoughtful curiosity about the nature of the universe. When the lunar module ***Antares*** *touched down on the Moon's* ***Fra Mauro*** *region on February 5, 1971, Mitchell and Shepard stepped onto the surface of another world. Over the course of their mission, they* *conducted experiments, collected lunar samples, and explored terrain that had once been the planned destination of Apollo 13.*

But for Mitchell, the most profound experience occurred not on the lunar surface—but during the journey home. As the spacecraft traveled through the vast darkness between the Moon and Earth, Mitchell gazed out the window and contemplated the universe. Seeing the Earth suspended in space inspired a deep sense of connection with the cosmos.

The experience profoundly affected him. After leaving NASA, Mitchell devoted much of his life to exploring philosophical and scientific questions about consciousness, the universe, and humanity's place within it. He founded the ***Institute of Noetic Sciences****, an organization dedicated to studying the relationship between human awareness and the broader universe.*

Edgar Mitchell was not only a lunar explorer but also a thinker who sought to understand the deeper meaning of humanity's journey into space. His reflections remind us that space exploration is not only about technology and discovery. It is also about understanding our place in the universe.

Edgar Dean Mitchell (September 17, 1930 – February 4, 2016) was a United States Navy officer and aviator, test pilot, aeronautical engineer, ufologist, and NASA astronaut. As the Lunar Module Pilot of Apollo 14 in 1971 he spent nine hours working on the lunar surface in the Fra Mauro Highlands region, and was the sixth person to walk on the Moon. He was the second Freemason to set foot on the Moon, after Buzz Aldrin.

---

1. Dr. Mitchell, what first inspired your interest in science and exploration?

From an early age I was fascinated by science and the mysteries of the universe. Growing up in New Mexico, I often looked at the night sky and wondered about the vast cosmos beyond Earth. That curiosity eventually led me to study engineering and later become involved in the space program.

---

2. What did it mean to you to be selected as an astronaut for the Apollo program?

The Apollo program represented one of humanity's greatest scientific endeavors. Being selected as an astronaut was both an honor and a tremendous responsibility. We were not only flying missions, but we were also helping expand human knowledge.

---

3. Dr. Mitchell, what do you remember about the landing of the lunar module Antares on the Moon?

The landing of the lunar module *Antares* during the Apollo 14 Moon Landing was one of the most intense and memorable moments of the entire mission. As we descended toward the Fra Mauro region, every second required careful concentration.

Inside the lunar module, Commander Alan Shepard and I closely monitored the instruments—our altitude, descent rate, and the computer guidance system. At one point, we encountered a technical issue with the landing guidance that briefly threatened to abort the mission. Mission Control worked rapidly with us to reset the system and confirm that we could safely continue the descent.

As the surface of the Moon came into clearer view through the windows, the landscape appeared stark and dramatic—craters, ridges, and long shadows stretching across the gray terrain. There is no atmosphere on the Moon, so the light is incredibly sharp and the horizon appears closer than it does on Earth.

When the engine finally shut down and we heard the words confirming that we had landed, there was a deep sense of relief and accomplishment. After years of preparation, we had safely arrived on another world.

Looking out the window of *Antares*, I remember thinking about the long journey that had brought humanity to that moment—from the earliest rockets to the missions that carried us across nearly a quarter million miles of space. It was a powerful reminder that exploration is part of our nature, and that the Moon was only the beginning of humanity's expansion into the universe. 🌙🚀

Mitchell stands with the United States' flag on the Moon, 1971, Source/Auhor: NASA – Public Domain

---

4. What was your first impression when you stepped onto the Moon?

When I stepped onto the Moon during the mission of Apollo 14 Moon Landing, my first impression was the extraordinary clarity of the landscape. The lunar surface appeared stark and ancient—craters, rocks, and fine gray dust stretching across the Fra Mauro region.

Because the Moon has no atmosphere, the sky above was completely black even though the Sun was shining brightly. The light created sharp contrasts between illuminated areas and deep shadows, giving the terrain an almost surreal appearance.

Another striking sensation was the Moon's gravity. At only about one-sixth that of Earth, moving across the surface felt almost effortless. With each step you could bound lightly across the terrain, which made walking feel both unfamiliar and exhilarating.

Mitchell studies a map while walking on the Moon, February 6, 1971, Source/Author: NASA Alan Shepard – Public Domain

But perhaps the most powerful impression was the realization that we were standing on another world. As a scientist and explorer, I was deeply aware that the rocks and soil beneath our boots contained clues about the early history of the Moon—and possibly the early history of the entire solar system.

At that moment, I felt both excitement and humility. We were explorers representing not just a nation, but humanity itself, conducting scientific discovery on a world that had remained untouched for billions of years.

---

5. During Apollo 14 you conducted important scientific experiments. Why were these experiments significant?

By the time of the Apollo 14 Moon Landing, the Apollo program had evolved from simply demonstrating that humans could reach the Moon to conducting serious scientific exploration. Our mission focused heavily on understanding the Moon's geology and its history.

One of our main goals was to study the Fra Mauro region, which scientists believed contained material ejected from the enormous impact that formed the Imbrium Basin billions of years ago. By collecting rock and soil samples from this area, we hoped to learn more about the early history of the Moon and the violent events that shaped the solar system.

We also deployed a series of instruments known as the Apollo Lunar Surface Experiments Package (ALSEP). These instruments measured seismic activity, detected tiny moonquakes, studied the solar wind, and monitored the Moon's internal structure. The data from these instruments continued to be transmitted back to Earth long after we had left the lunar surface.

The significance of these experiments lies in the fact that the Moon preserves an ancient record of planetary history. Unlike Earth, where weather and geological activity constantly reshape the surface, the Moon has remained largely unchanged for billions of years. In that sense, the scientific work we conducted on the Moon helped humanity better understand not only the Moon itself, but also the origins and evolution of the Earth and the entire solar system. 🌜🚀

A U.S. Navy diver helps Mitchell out of the Command Module, 1971, National Museum of the US Navy – Public Domain

---

6. During the return journey to Earth, you experienced what you later described as a profound realization. Could you explain that moment?

Yes, that moment occurred during our journey home from the Apollo 14 Moon Landing. As the spacecraft traveled through the vast darkness between the Moon and Earth, I spent time looking out the window at our planet.

From that distance, Earth appeared as a brilliant blue and white sphere floating in space. It was stunningly beautiful, but also strikingly small. In that moment, I experienced something that is difficult to describe fully—a deep sense of unity and connection with the entire planet.

I suddenly realized that everything humanity has ever known—every culture, every civilization, every human story—exists on that small world suspended in the cosmos. There were no borders visible, no divisions between nations, only one shared home.

The experience produced what I later described as a kind of "overview effect," a powerful awareness of the interconnectedness of all life on Earth. It was both emotional and intellectual at the same time.

That realization stayed with me long after I returned home. It reinforced my belief that humanity must learn to see itself as part of a single planetary community, working together to protect and understand the remarkable world we inhabit. 🌍🚀

---

7. How did that experience influence your life after leaving NASA?

The realization I experienced during the return from the Apollo 14 Moon Landing had a lasting influence on the direction of my life after leaving NASA. Seeing Earth from deep space gave me a profound sense of the interconnectedness of all life. It was not only an intellectual insight but also a deeply personal experience that made me reflect on humanity's place in the universe.

After my career as an astronaut, I became increasingly interested in the relationship between science, human consciousness, and the broader understanding of reality. I felt that the perspective gained from space exploration could inspire new ways of thinking about humanity's future and our responsibility toward our planet.

In 1973, I founded the Institute of Noetic Sciences, an organization dedicated to exploring questions about consciousness, human potential, and the nature of awareness. My goal was to encourage scientific inquiry into areas that connect physical sciences with human experience.

The journey to the Moon taught me that exploration is not only about discovering new worlds in space, but also about expanding our understanding of ourselves. In that sense, my work after NASA was simply another form of exploration, one focused on the inner dimensions of human knowledge and consciousness. 🌍🚀

---

8. What do you believe is the most important lesson humanity can learn from space exploration?

Space exploration helps us understand that we are part of a much larger universe. It encourages us to think beyond national boundaries and to recognize our shared responsibility for the planet.

---

9. How do you see the future of human exploration in space?

I believe the future of human exploration in space will unfold in stages, gradually expanding humanity's presence beyond Earth. The Moon will likely serve as the first major stepping stone—a place where we can build research stations, test technologies, and learn how to live and work on another world for extended periods of time.

From there, the next great objective will be Mars. Sending humans to Mars will require new propulsion systems, advanced life-support technologies, and a level of international cooperation even greater than what we saw during the Apollo era. But the challenge is well within humanity's capabilities.

Beyond the technical achievements, space exploration also carries a deeper meaning. When astronauts travel beyond Earth and look back at our planet, they often experience a powerful sense of unity and perspective. That realization can influence how we think about our responsibilities as inhabitants of a single world.

I believe that as humanity expands into space, we will not only discover new environments and new scientific knowledge—we will also gain a better understanding of ourselves. Exploration pushes the boundaries of human imagination, and in doing so it encourages cooperation, innovation, and a broader sense of planetary identity.

The Moon and Mars are not the end of the journey. They are simply the first steps in humanity's long adventure into the cosmos. 🚀

Edgar Mitchell's memorial in the Treasures of Apollo exhibit at Kennedy Space Center, Source/Author: NASA/Bill White – Public Domain

---

10. What message would you share with future generations looking toward the stars?

My message to future generations is simple: never lose your sense of wonder. Human progress has always been driven by curiosity, the desire to ask questions about the unknown and to explore what lies beyond the horizon. When we traveled to the Moon during the Apollo 14 Moon Landing, we were continuing a tradition of exploration that began long before the space age.

The universe is vast, and humanity has only begun to explore it. Future generations will build lunar bases, travel to Mars, and perhaps one day journey to distant worlds beyond our solar system.

But as you reach the stars, remember to look back at Earth. From space, our planet appears as a small, beautiful oasis in an immense cosmic ocean. It reminds us that all life on Earth shares one home.

Let exploration inspire not only technological progress, but also wisdom and responsibility. The greatest discoveries we make in space may ultimately help us better understand ourselves and our place in the universe. The stars have always guided humanity. The next chapter of that journey belongs to you. 🌍🚀✨

Interview with Gene Cernan

Commander of Apollo 17

Introduction

In December 1972, humanity completed one of the most remarkable chapters in the history of exploration. For several years, astronauts had traveled nearly 400,000 kilometers from Earth to the Moon, landing on its silent surface and returning with knowledge that transformed science. Each mission expanded our understanding of the Moon and demonstrated the extraordinary capabilities of human exploration.

The final mission of this era was ***Apollo 17****. Commanding the mission was astronaut Gene Cernan, a naval aviator known for his determination, skill, and deep appreciation for the significance of space exploration. Alongside astronaut Harrison Schmitt and command module pilot Ronald Evans, Cernan prepared for what would become the last human journey to the Moon of the twentieth century.*

When the lunar module ***Challenger*** *touched down in the Taurus–Littrow Valley on December 11, 1972, Cernan and Schmitt stepped onto a landscape surrounded by towering* *mountains and ancient volcanic formations. Using the lunar rover, they explored one of the most geologically interesting sites ever visited on the Moon. Over three days, they conducted experiments, collected lunar samples, and traveled farther across the lunar surface than any astronaut before them.*

Yet as the mission ended, a moment of quiet reflection arrived. Standing beside the lunar module, Gene Cernan prepared to climb the ladder that would carry him back into the spacecraft and ultimately back to Earth. Before leaving the Moon, he paused and looked once more across the silent valley.

His footprints in the lunar dust would remain there for millions of years. Cernan knew that when he left the Moon, no other human had plans to return soon. With those thoughts in mind, he spoke words that captured both the pride of the Apollo era and the hope that humanity would someday return:

"We leave as we came and, God willing, as we shall return—with peace and hope for all mankind." With that final step, Gene Cernan became the ***last human being to walk on the Moon****. More than half a century later, his footsteps remain undisturbed in the lunar soil—silent reminders of humanity's greatest journey beyond Earth.*

Eugene Andrew Cernan (March 14, 1934 – January 16, 2017) was an American astronaut, naval aviator, electrical engineer, aeronautical engineer, and fighter pilot. Cernan traveled into space three times and to the Moon twice: as pilot of Gemini 9A in June 1966, as Lunar Module pilot of Apollo 10 in May 1969, and as commander of Apollo 17 in December 1972, the final Apollo lunar landing. He was also a backup crew member of Gemini 12, Apollo 7, and Apollo 14. During the Apollo 17 mission, he became the 11th person to walk on the Moon as well as the astronaut with the longest time spent walking on the Moon; he is also the last person to walk on the Moon.

1. Captain Cernan, when did your interest in aviation and exploration begin?

My interest in aviation began during my youth when I became fascinated by airplanes and flight. Eventually I attended Purdue University and later became a naval aviator. Flying high-performance aircraft prepared me well for the challenges of becoming an astronaut.

---

2. What did it mean to you to be selected as an astronaut during the Apollo era?

It was an incredible privilege. The Apollo program represented humanity's boldest exploration effort, and everyone involved understood that we were participating in something that would be remembered for generations.

---

3. Before Apollo 17, you flew on Gemini 9 and Apollo 10. How did those missions prepare you for commanding a lunar landing?

Each mission I flew before commanding Apollo 17 Moon Landing was an important step in preparing me for that responsibility. My first spaceflight was the Gemini 9A mission, where I served as pilot. During that mission, I performed a spacewalk that proved to be far more difficult than anyone had anticipated. The experience taught me valuable lessons about working in space, the limits of equipment, and the importance of preparation and teamwork.

Then came Apollo 10 mission, which was essentially the full-dress rehearsal for the first Moon landing. As the lunar module pilot, I flew the lunar module *Snoopy* with Commander Tom Stafford down to within about nine miles of the Moon's surface. That mission tested nearly every step required for a landing—navigation, descent procedures, and rendezvous with the command module.

Apollo 10 gave us a detailed understanding of how the spacecraft behaved in lunar orbit and during descent. It also gave me direct experience flying the lunar module, which was critical preparation for commanding a landing mission.

By the time Apollo 17 came along, I had experienced both the challenges of working outside a spacecraft and the complex operations required around the Moon. Those earlier missions provided the knowledge, confidence, and perspective needed to lead a mission that would place astronauts on the lunar surface. Space exploration is built step by step, and every mission builds on the lessons learned from the ones before it.

Cernan aboard Gemini 9A, Source/Author: NASA – Public Domain

---

4. What was your first impression when you landed in the Taurus–Littrow Valley?

When we landed in the Taurus–Littrow Valley during the Apollo 17 Moon Landing, my first impression was how dramatic and magnificent the landscape appeared. Unlike the relatively flat landing sites of earlier missions, Taurus–Littrow was surrounded by towering mountains that rose thousands of feet above the valley floor.

Looking out the window of the lunar module *Challenger*, I remember seeing the massive South Massif and North Massif rising sharply against the black sky. The valley itself stretched out before us like a vast natural amphitheater carved into the lunar surface.

The scene was both beautiful and humbling. The mountains were ancient, formed billions of years ago, and the surface around us had remained largely unchanged since that time. Standing there felt like stepping into a geological record of the early solar system.

There was also a tremendous sense of excitement. The site had been chosen because scientists believed it contained materials from both volcanic activity and deep lunar impacts. That meant we had an extraordinary opportunity to collect samples that could reveal important clues about the Moon's history.

Cernan the Moon during Apollo 17 (December 13, 1972)
Source/Author: NASA/Harrison H. Schmitt – Public Domain

In that moment, I realized that Apollo 17 was not only about exploration—it was about discovery. We were scientists as well as astronauts, exploring one of the most fascinating landscapes ever visited by humans. 🌙🚀

---

5. Apollo 17 included the first professional geologist to walk on the Moon, Harrison Schmitt. How did that influence the mission?

Having Harrison Schmitt on the crew during the Apollo 17 Moon Landing made a tremendous difference in the scientific value of the mission. Earlier Apollo missions had already begun focusing more on geology, but Harrison was a trained field geologist whose entire career had been devoted to studying planetary surfaces. His ability to quickly recognize significant geological features and identify important samples was invaluable.

When we explored the Taurus–Littrow Valley, Harrison could immediately interpret what we were seeing—layers of rock, volcanic deposits, and impact materials that told the story of the Moon's past. His observations helped guide our sampling decisions so that we could collect the most scientifically valuable materials.

One of the most exciting moments came when Harrison discovered the famous "orange soil." That unusual material turned out to be evidence of ancient volcanic activity on the

Moon, providing important clues about the Moon's geological history.

Working with a scientist like Harrison reminded us that Apollo was not only an engineering achievement but also a scientific expedition. The rocks and samples we collected have continued to be studied for decades, helping researchers better understand the formation and evolution of the Moon and the early solar system. In many ways, Apollo 17 demonstrated what future planetary exploration should look like: astronauts and scientists working together to unlock the secrets of other worlds. 🌙🚀

---

6. What was it like driving the lunar rover across the Moon's surface?

Driving the lunar rover during the mission of Apollo 17 Moon Landing was one of the most exciting and useful parts of exploring the Moon. Before the rover was introduced, astronauts had to conduct most of their exploration on foot, which limited how far they could travel from the lunar module. The rover changed everything. It allowed us to travel several kilometers across the Taurus–Littrow Valley and reach geological sites that would have been impossible to visit otherwise.

The experience itself was unlike driving anything on Earth. The Moon's gravity is only about one-sixth that of Earth, so the rover felt very light and bounced easily over the terrain. When we drove across small craters or rocks, the vehicle would sometimes lift slightly off the surface before settling back down.

Despite the rough terrain, the rover performed remarkably well. It was simple, reliable, and designed perfectly for the lunar environment. With it, we were able to carry scientific instruments, collect samples, and explore much larger areas of the valley.

But perhaps the most memorable part was the view. As we drove across the gray lunar landscape with the towering mountains of Taurus–Littrow surrounding us and the black sky above, it felt like traveling through a world that had remained unchanged for billions of years. In those moments, it truly felt like humanity had become a species capable of exploring other worlds. 🚀🌙

---

7. What moment during the mission stands out most in your memory?

There were many unforgettable moments during the mission of Apollo 17 Moon Landing, but the one that stands out most clearly in my memory occurred near the very end of our time on the Moon.

After completing our exploration of the Taurus–Littrow Valley and collecting an extraordinary number of geological samples, it was time to return to the lunar module *Challenger*.

As commander, I was the last astronaut to climb the ladder and re-enter the spacecraft.

Before leaving the surface, I paused for a moment and looked out across the lunar landscape. The valley was quiet and unchanged, just as it had been for billions of years. In that moment, I realized that I might be the last human being to walk on the Moon for a long time.

I wanted to leave a message that captured both the spirit of exploration and the hope that humanity would continue the journey. As I said before climbing the ladder, we were leaving the Moon "as we came and, God willing, as we shall return—with peace and hope for all mankind."

That moment represented not only the end of our mission but also the closing chapter of the Apollo program. Yet I never believed those were humanity's final footsteps on the Moon, the most recent ones in a much longer story of exploration. 🌙🚀

---

8. Commander Cernan, during the mission of *Apollo 17 Moon Landing* you became the last human to leave the lunar surface. How did you feel when you realized you would be the final person to step off the Moon?

At the time, I was very aware that I might be the last person to walk on the Moon for quite some time. As commander of Apollo 17, I had the privilege of being the final astronaut to climb back into the lunar module after our exploration.

Standing there on the lunar surface, I felt a mixture of pride, humility, and a little sadness. Pride, because our mission represented the culmination of the entire Apollo program. Humility, because I knew that what we accomplished was made possible by hundreds of thousands of people who had worked for years to make those missions possible.

But there was also a sense of unfinished business. As I looked out across the Moon's landscape, I was certain that humanity would return one day. The Moon still had many secrets to reveal, and our visit there was only the beginning.

Before climbing the ladder of the lunar module *Challenger*, I tried to leave a message that captured that feeling. I said that we were leaving the Moon "as we came and, God willing, as we shall return, with peace and hope for all mankind." I never thought of myself as the *last* person on the Moon—only as the most recent one. Exploration is part of human nature, and sooner or later, others will follow those footprints again.

---

9. What do you believe was the greatest achievement of the Apollo program?

Most people would say the greatest achievement of the Apollo program was landing humans on the Moon. And of course, when missions like Apollo 12 Moon Landing successfully landed astronauts on the lunar surface, it marked one of the most dramatic milestones in human history.

Cernan's lunar space suit on display at the National Air and Space Museum in Washington, D.C., Author Tyrol5 – PD

But from my perspective, the true achievement went far beyond the Moon itself. The Apollo program proved that an enormous and complex goal could be achieved when thousands of people worked together with determination and discipline. Engineers, scientists, technicians, and astronauts across the country collaborated to solve problems that had never been faced before. They developed powerful rockets, advanced spacecraft, and new methods of navigation and communication.

In a very real sense, Apollo showed what humanity is capable of when it commits itself to exploration and innovation. The Moon landings were the visible result, but the deeper accomplishment was the spirit of cooperation and ingenuity that made them possible. That spirit pushed the boundaries of science and technology and inspired generations of young people to pursue careers in engineering, physics, and space exploration.

In the end, Apollo was not just about reaching another world, it was about proving that human imagination and determination can turn the impossible into reality.

---

10. After your remarkable career at *NASA*, you transitioned into leadership roles in several major companies. Could you tell us more about this phase of your life and what drew you into the business world?

After completing my missions with NASA—including commanding the mission of Apollo 17 Moon Landing, the final Apollo landing—I felt that it was time to begin a new chapter in my life. The space program had given me extraordinary experiences, but it had also taught me valuable skills that were just as important outside the astronaut corps: leadership, decision-making under pressure, teamwork, and managing complex technological projects. Those skills translated naturally into the world of business.

I became involved with several companies in the aerospace and energy industries, working in executive and consulting roles. These positions allowed me to continue contributing to technological innovation and to help guide organizations facing complicated engineering and operational challenges.

What attracted me most to the business world was the opportunity to keep building new projects, new partnerships, and new ideas. In many ways, it reminded me of the early days of the space program, when teams of engineers and scientists were trying to turn ambitious visions into reality.

At the same time, I spent a great deal of time speaking publicly about the importance of exploration, education, and scientific progress. I felt a responsibility to share the lessons learned during the Apollo era and to encourage future generations to pursue careers in science, engineering, and space exploration. Leaving NASA did not mean leaving exploration behind. It simply meant continuing that spirit of discovery in new arenas here on Earth. 🚀

---

11. What message would you share with future explorers who will return to the Moon or travel to Mars?

To the explorers who will follow those who will return to the Moon and eventually journey to Mars, I would say this: you are continuing one of humanity's greatest adventures.

When we left the Moon during the mission of Apollo 17 Moon Landing, I believed strongly that we were not closing a door but opening one. The footprints we left in the lunar dust were meant to mark the beginning of humanity's exploration beyond Earth.

The challenges you will face will be greater than ours. Missions to Mars will require longer voyages, new technologies, and an extraordinary level of teamwork and determination. But exploration has always demanded courage and perseverance.

As you travel to other worlds, remember the perspective that space gives us. When you look back at Earth from the Moon or from deep space, you will see a small, beautiful planet without borders—a reminder that all humanity shares one home.

Carry that understanding with you. Explore with curiosity, humility, and a spirit of cooperation. The future of space exploration will belong not just to one nation but to all people working together. And when you set foot on new worlds, remember that you are not only explorers—you are ambassadors for all of humanity. 🚀🌍✨

Interview with Jim Lovell

Commander of Apollo 13

Introduction

In April 1970, humanity prepared for another mission to the Moon. The Apollo program had already achieved several successful lunar landings, and Apollo 13 was expected to continue that remarkable series of explorations. But space flight has always carried risks.

Commanding the mission was astronaut Jim Lovell, one of NASA's most experienced astronauts and a veteran of multiple space missions. Lovell had already flown in space three times, including ***Apollo 8****, the historic mission that first carried humans into orbit around the Moon.*

Apollo 13 seemed destined to be another successful journey to the lunar surface. Yet two days after launch, as the spacecraft traveled nearly 320,000 kilometers from Earth, a sudden explosion shook the command module ***Odyssey****. Warning lights flashed across the control panels as oxygen tanks ruptured and critical systems began shutting down.*

Lovell calmly transmitted a message to Mission Control that would become one of the most famous phrases in the history of space exploration: "Houston, we've had a problem." The explosion crippled the spacecraft and forced NASA to abandon the planned lunar landing. Instead, the mission became a desperate struggle to bring the three astronauts safely back to Earth.

With power and oxygen rapidly diminishing, Lovell and his crew moved into the lunar module ***Aquarius****, using it as a lifeboat while engineers on Earth raced to devise solutions. Over the next four days, astronauts and ground controllers worked together under immense pressure to overcome one crisis after another. Improvised repairs, carefully calculated maneuvers, and extraordinary teamwork allowed the spacecraft to navigate around the Moon and begin its journey home.*

Against overwhelming odds, Apollo 13 returned safely to Earth. Though the mission never landed on the Moon, it became one of NASA's greatest demonstrations of ingenuity and courage. Jim Lovell's calm leadership and determination helped guide the crew through one of the most dangerous situations ever faced in human spaceflight. Today, Apollo 13 is remembered not as a failure, but as a triumph of human resilience.

James Arthur Lovell Jr. (March 25, 1928 – August 7, 2025) was an American astronaut, naval aviator, test pilot, and mechanical engineer. In 1968, as command module pilot of Apollo 8, he, along with Frank Borman and William Anders, became one of the first three astronauts to fly to and orbit the Moon. He then commanded the Apollo 13 lunar mission in 1970 which, after a critical failure en route, looped around the Moon and returned safely to Earth.

---

1. Captain Lovell, when did your fascination with aviation and exploration begin?

My interest in aviation began during my youth when I became fascinated by airplanes and flight. Eventually I joined the U.S. Navy and trained as a naval aviator. Flying high-performance aircraft naturally led me toward the opportunity to become an astronaut.

---

2. You flew on several historic missions before Apollo 13. What was your most memorable experience prior to that mission?

Apollo 8 was an extraordinary experience. It was the first mission to carry humans into lunar orbit. Seeing the Earth rise above the Moon's horizon was one of the most beautiful sights I have ever witnessed.

---

3. Captain Lovell, what do you remember about the moment when the explosion occurred aboard Apollo 13?

The moment of the explosion during the mission of Apollo 13 mission was sudden and completely unexpected. We were about two days into the flight, conducting what seemed like a routine procedure—stirring the oxygen tanks in the service module.

Almost immediately after the switch was activated, there was a loud bang followed by a noticeable shudder through the spacecraft. At first it felt as if something had struck us from the outside. Warning lights began to appear on the control panel, and several instruments started showing abnormal readings.

Very quickly we realized that something serious had happened. Oxygen levels were dropping, and the spacecraft's electrical systems were becoming unstable. That was when Jack Swigert reported to Mission Control the now-famous words: *"Houston, we've had a problem."*

Lovell (second from left) and Buzz Aldrin (third from left) arrive aboard the recovery vessel, the aircraft carrier USS Wasp, after the Gemini 12 flight, Source/Author: *NASA – Public Domain*

In those first moments, our focus shifted entirely to understanding what had gone wrong and how we could stabilize the spacecraft. There was no time for panic—only for careful observation and disciplined problem-solving.

Looking back, that moment marked the beginning of one of the most extraordinary rescue efforts in the history of space exploration. What followed was an incredible example of teamwork between the astronauts and the engineers at NASA, working together to bring the crew safely back to Earth.

---

4. When you realized the Moon landing would have to be aborted, what went through your mind?

When the explosion occurred during the mission of Apollo 13 mission, our first concern was not the Moon landing—it was survival. At first, we were trying to understand what had happened. As the situation became clearer, we realized that the oxygen tanks in the service module had been severely damaged and that the command module was rapidly losing both oxygen and electrical power. Without those systems, it would be impossible to continue the mission to the Moon.

The realization that the lunar landing had to be aborted came quickly and almost naturally. In that moment, there was no time to dwell on disappointment. Our focus shifted immediately to the problem of how to keep the crew alive and find a way back to Earth.

Of course, there was a sense of sadness. As astronauts, we had trained for years with the dream of landing on the Moon. But space exploration always carries risks, and we understood that the mission had changed from exploration to survival.

From that point forward, every decision was focused on one goal: conserving power, preserving oxygen, and working with the incredible team at NASA to safely return home. In the end, Apollo 13 did not land on the Moon, but it became something else—a powerful demonstration of human ingenuity, teamwork, and determination in the face of a life-threatening challenge. 🚀

Lovell at the Command Module Guidance and Navigation station during the Apollo 8 mission

---

5. How did the lunar module become the crew's lifeboat?

The lunar module was designed to support two astronauts on the Moon for a limited time. We adapted it to support three astronauts during the journey back to Earth. It required careful management of power, oxygen, and other resources.

---

6. What role did Mission Control play in solving the crisis?

Mission Control played an essential role. Engineers on the ground worked continuously to develop procedures that would allow us to conserve power, remove carbon dioxide from the cabin, and navigate safely home.

---

7. One of the most famous challenges was adapting square carbon dioxide filters to fit round openings. How did you handle that situation?

Engineers on the ground designed an improvised solution using materials available inside the spacecraft. Following their instructions, we assembled the device, which successfully reduced carbon dioxide levels inside the cabin.

---

8. What was the most difficult moment during the mission?

The mission of Apollo 13 had several extremely challenging moments, but one of the most difficult came after we understood the full extent of the damage to the spacecraft.

The explosion had crippled the service module of the command module *Odyssey*. Our oxygen supply was rapidly disappearing, and the electrical systems that powered the spacecraft were failing. Without those systems, we would not be able to survive long enough to return to Earth.

At that point we had to move into the lunar module *Aquarius*, which had originally been designed to support two astronauts for only about two days on the Moon. Now it had to serve as a lifeboat for three astronauts for nearly four days.

Another major challenge was the buildup of carbon dioxide inside the spacecraft. The filters in the lunar module were not designed for three people over such a long period. Engineers at NASA worked tirelessly to design an improvised solution using materials available aboard the spacecraft. Following their instructions, we assembled the device in orbit and successfully reduced carbon dioxide levels.

Throughout the mission, every step required careful planning and close cooperation between the crew and Mission Control. The difficulties were significant, but the teamwork and ingenuity involved ultimately allowed us to bring the spacecraft safely back to Earth.

In many ways, the most difficult moments also demonstrated the greatest strength of the Apollo program—the ability of people to work together and solve problems even in the most dangerous situations.

---

9. How did the successful return of Apollo 13 affect your view of space exploration?

Apollo 13 demonstrated the incredible teamwork between astronauts and engineers. The mission showed that even in the face of serious adversity, determination and ingenuity can overcome enormous challenges.

Group photo of NASA Astronaut Group 2; Lovell is third from the left in the back row, Source/Author: NASA – Public Domain

---

10.In a 2020 episode of *Houston We Have a Podcast*, you said: "You can't suddenly have a problem and then just close your eyes and hope for a miracle, because a miracle is something you have to make happen yourself—or with the help of others." Could you explain what you meant by that?

What I meant by that statement comes directly from the lessons we learned during the Apollo program at NASA. Spaceflight is an environment where problems can appear suddenly and where there is very little margin for error. When something goes wrong, you cannot rely on luck or simply hope that the situation will fix itself. Instead, you must rely on preparation, knowledge, teamwork, and calm decision-making.

Throughout the Apollo missions, including the mission of Apollo 17 Moon Landing—astronauts and engineers trained constantly for unexpected situations. We practiced emergencies again so that when real problems appeared, we were ready to respond.

In that sense, the "miracle" people sometimes talk about is the result of human effort. It is created by the engineers who design the systems, the flight controllers who monitor every detail, and the astronauts who remain disciplined under pressure. No one accomplishes something like space exploration alone. It requires thousands of people working together, each doing their part with dedication and professionalism.

So, when I say that a miracle is something you make yourself, what I really mean is that success comes from preparation, teamwork, and the willingness to confront challenges directly rather than hoping they will disappear. That philosophy applies not only to spaceflight, but to many challenges in life. 🚀

---

11. What message would you share with future astronauts exploring the Moon and Mars?

Exploration requires preparation, teamwork, and resilience. Space missions will always involve challenges, but those challenges also inspire innovation and discovery. Future explorers will continue pushing the boundaries of what humanity can achieve.

Lovell (left), Gene Cernan (center), and Neil Armstrong during the 2010 Legends of Aerospace tour aboard the aircraft carrier USS Harry S. Truman U.S. Navy photo by Mass Communication Specialist 2nd Class Kilho Park - Public Domain

Interview with Roger Chaffee

Apollo 1 Astronaut

Introduction

The early years of space exploration were filled with ambition, discovery, and extraordinary courage. Yet they also carried profound risks. Every rocket launch, every spacecraft design, and every mission represented a leap into the unknown. Among the astronauts chosen to lead America's journey to the Moon was a young naval officer and engineer named Roger Bruce Chaffee.

Born in Michigan and trained as a naval aviator, Chaffee quickly distinguished himself through intelligence, discipline, and exceptional technical skills. His calm demeanor and sharp engineering mind earned him a place among NASA's third group of astronauts, selected in 1963 to support the rapidly expanding Apollo program.

Although he had not yet flown in space, Chaffee played an important role in developing procedures and testing systems that would guide future missions. In 1967, he was assigned as the lunar module pilot for ***Apollo 1****, the first crewed mission of the Apollo program. The mission represented a* *critical step toward NASA's goal—landing humans on the Moon.*

Chaffee's crewmates were two of NASA's most experienced astronauts: commander Gus Grissom and senior pilot Ed White, the first American to perform a spacewalk. On January 27, 1967, the crew entered the Apollo command module for a routine ground test at Cape Kennedy (now Cape Canaveral). The test was intended to simulate launch conditions while the spacecraft remained on the launch pad.

Suddenly, a spark ignited inside the cabin. Within seconds, the oxygen-rich atmosphere inside the spacecraft fueled a catastrophic fire. The flames spread rapidly through the capsule, and despite desperate efforts by the crew and ground personnel, Grissom, White, and Chaffee lost their lives.

The tragedy shocked NASA and the world. Yet from this devastating loss came crucial lessons. Engineers conducted a thorough investigation that led to major improvements in spacecraft design, materials, wiring, and emergency procedures. These changes dramatically increased the safety of the Apollo spacecraft.

Just two years later, astronauts successfully land on the Moon. Roger Chaffee never had the chance to fly into space, but his dedication and sacrifice helped make the achievements of Apollo possible. His legacy lives on in every mission that followed. The path to the Moon was built not only by triumphs—but also by those who gave their lives in pursuit of exploration.

Roger Bruce Chaffee (February 15, 1935 – January 27, 1967) was an American naval officer, aviator and aeronautical engineer who was a NASA astronaut in the Apollo program.

---

1. Lieutenant Commander Chaffee, what first inspired your interest in aviation and engineering?

From a young age I was fascinated by both flight and technology. Aviation combined the excitement of exploration with the challenge of engineering. That interest eventually led me to pursue a career as a naval aviator and later to join NASA.

---

2. How did you feel when you were selected as an astronaut in NASA's third group?

Being selected as an astronaut was an incredible honor. The Apollo program represented one of the most ambitious scientific and engineering efforts ever attempted, and I was excited to contribute to its success.

---

3. Although Apollo 1 would have been your first spaceflight, you were already deeply involved in mission preparation. What did that work involve?

Much of our work focused on testing spacecraft systems, developing procedures, and ensuring that astronauts could operate the Apollo spacecraft safely and effectively during missions.

---

4. What qualities do you believe are most important for astronauts?

I believe that the most important qualities for astronauts begin with **discipline and dedication**. Spaceflight is an extremely demanding profession where even the smallest detail can make a difference. Astronauts must be willing to train constantly, learn complex systems, and approach every task with complete focus.

Another essential quality is **teamwork**. No astronaut ever accomplishes anything alone. Every mission is supported by thousands of engineers, technicians, and scientists working behind the scenes at organizations like NASA. Trust between the crew and the people on the ground is absolutely critical.

Courage is also important, but courage does not mean ignoring risk. It means understanding the risks and accepting them in the pursuit of exploration and scientific discovery. Space exploration has always involved challenges, and astronauts must be prepared to face them with calm determination.

Finally, astronauts need a deep sense of **curiosity and purpose**. The goal of spaceflight is not only to travel beyond Earth but to expand human knowledge. Those who choose this path must believe strongly in the value of exploration and in the idea that humanity's future may extend far beyond our home planet.

In the end, astronauts are simply people who combine skill, preparation, and teamwork with the courage to explore the unknown. 🚀

Chaffee (sitting on hatch sill) during water egress training for Apollo 1

5. What excited you most about the Apollo program?

Apollo represented humanity's opportunity to explore beyond Earth in a way never attempted before. Landing on the Moon was not only a technological achievement, but it was also a symbol of human curiosity and determination.

6. How did astronauts prepare for the challenges of spaceflight?

Preparing for spaceflight required an enormous amount of training and discipline. At NASA, astronauts spent countless hours learning every system of the spacecraft in detail—electrical systems, navigation, communications, and life-support equipment. We had to understand not only how the spacecraft worked, but also how to respond if something went wrong.

One of the most important aspects of preparation was **simulation training**. Astronauts spent many hours in spacecraft simulators that recreated the exact environment of the capsule. In these simulations, instructors would introduce unexpected problems so that the crew had to learn how to respond quickly and calmly.

Physical training was also essential. Astronauts trained in centrifuges to experience the intense forces of launch and reentry, and they practiced working in conditions that simulated weightlessness. For spacewalk preparation, we often trained underwater, which helped reproduce the feeling of moving in microgravity.

Equally important was **team coordination**. Astronauts trained closely with flight controllers and engineers so that everyone understood the mission procedures. Spaceflight depends on the combined efforts of the crew and the thousands of people supporting them on the ground.

In the end, preparation was the key to success. Space exploration leaves very little room for improvisation. The better prepared the crew is, the more effectively they can respond when unexpected challenges arise.

Chaffee at the consoles in Mission Control during the Gemini 3 mission, Source/Author: NASA – Public Domain

---

7. What do you believe makes space exploration worth the risks involved?

Space exploration has always involved risk, but throughout history humanity has accepted risks to expand knowledge and explore new frontiers. The exploration of space continues that tradition.

One of the most important reasons space explorations is worth the risk is scientific discovery. By traveling beyond Earth, we gain new knowledge about the Moon, the planets, and the origins of the solar system. That knowledge not only advances science but also deepens our understanding of our own planet.

Another reason is technological progress. Programs at NASA have driven innovations in engineering, communications, computing, and many other fields. Technologies developed for space missions often find applications that benefit people on Earth.

Perhaps most importantly, space exploration inspires humanity. When people see astronauts traveling beyond Earth, it reminds them that human beings can achieve extraordinary things through curiosity, teamwork, and determination.

Exploration has always been part of the human spirit. From the first voyages across oceans to the missions of the Apollo era, progress has come from those willing to venture into the unknown. In that sense, the risks of space exploration are balanced by the tremendous rewards—knowledge, inspiration, and the possibility that humanity's future may extend far beyond our home planet. 🚀🌍

---

8. What did work alongside astronauts like Gus Grissom and Ed White mean to you?

Working with astronauts like Virgil and Ed White was both an honor and a tremendous learning experience.

Gus Grissom was one of the pioneers of the American space program. Having flown in the Mercury and Gemini programs, he possessed a deep understanding of spacecraft operations and the realities of spaceflight. He was a calm and highly capable commander who expected the best from everyone around him.

Ed White brought a different kind of inspiration. He had performed the first American spacewalk during the Gemini 4 mission, and his enthusiasm for space exploration was contagious. Ed had a remarkable sense of curiosity and optimism that energized the entire crew.

For me, being part of that team was an extraordinary opportunity. Although I had not yet flown in space, I was learning every day from two astronauts who had already helped shape the early history of human spaceflight at NASA.

What I valued most was the sense of teamwork and mutual trust that developed between us. Space missions require complete confidence in the people working beside you. Each member of the crew depends on the others not only for the success of the mission, but for their safety as well.

Working alongside Gus and Ed reinforced my belief that space exploration is never the achievement of one individual. It is always the result of a dedicated team striving together toward a shared goal. 🚀

---

9. What do you hope humanity achieves through space exploration?

I believe humanity will continue to explore beyond Earth, discovering new knowledge about the universe and expanding our understanding of science and technology.

10. What message would you share with future astronauts?

To the astronauts who will follow in the generations ahead, my message would be this: approach exploration with dedication, humility, and a deep sense of responsibility.

Spaceflight represents one of humanity's most demanding and meaningful endeavors. Every mission depends not only on the astronauts aboard the spacecraft but also on the thousands of people working behind the scenes at organizations like NASA. Respect that teamwork and never forget that you represent the efforts of an entire community of scientists, engineers, and explorers.

Preparation is essential. Train thoroughly, learn your systems in detail, and be ready to face unexpected challenges with calm determination. Space exploration requires discipline and constant learning. But at the same time, never lose your sense of wonder. When you look back at Earth from space, you will see a small and beautiful planet that belongs to all humanity. That perspective reminds us why exploration matters.

The journey into space is not only about reaching new destinations, it is about expanding human knowledge and inspiring future generations. The responsibility of astronauts is not just to explore, but to help humanity better understand its place in the universe. 🚀🌍✨

Statue of Chaffee in Grand Rapids, Michigan
Source: Wikimedia Commons – CC0

Interview with Sally Ride

First American Woman in Space

Introduction

By the late 1970s, space exploration had already transformed human understanding of the universe. Astronauts had traveled to the Moon, built increasingly sophisticated spacecraft, and expanded scientific research in orbit. Yet one barrier remained unbroken in the United States. Until then, every American astronaut who had flown into space had been a man.

That changed when NASA began recruiting a new generation of astronauts for the ***Space Shuttle program****. Unlike earlier programs that focused primarily on military test pilots, the shuttle era welcomed scientists, engineers, and specialists from diverse backgrounds. Among those selected in 1978 was a brilliant young physicist named Sally Ride.*

Born in California, Ride developed an early passion for science and exploration. She studied physics at Stanford University, earning a doctorate before joining NASA as part of the first astronaut class to include women. Five years later, she would make history.

On June 18, 1983, Ride launched aboard the Space Shuttle ***Challenger*** *on mission* ***STS-7****. As the shuttle soared into orbit, she became the* ***first American woman to travel into space****. Her achievement inspired millions of people around the world.*

During the mission, Ride operated the shuttle's robotic arm to deploy satellites and conduct scientific experiments, demonstrating the crucial role astronauts could play in managing complex technologies in orbit. But her impact extended far beyond a single mission.

After leaving NASA, Ride devoted much of her career to encouraging young people, especially girls, to pursue careers in science, technology, engineering, and mathematics. Through educational initiatives and public outreach, she helped inspire a new generation of scientists and explorers. Sally Ride's journey represented more than a historic first. It marked an important step toward a future in which space exploration belongs to everyone.

Sally Kristen Ride (May 26, 1951 – July 23, 2012) was an American astronaut and physicist. Born in Los Angeles, she joined NASA in 1978, and in 1983 became the first American woman and the third woman to fly in space, after cosmonauts Valentina Tereshkova in 1963 and Svetlana Savitskaya in 1982. She was the youngest American astronaut to have flown in space, having done so at the age of 32.

1. Dr. Ride, when did your interest in science and exploration begin?

My interest in science began during my childhood. I was always curious about how the world works. Physics eventually became my passion because it offered a way to explore fundamental questions about nature and the universe.

---

2. How did you first learn about the opportunity to become a NASA astronaut?

While I was a graduate student at Stanford, I saw a newspaper advertisement announcing that NASA was recruiting astronauts—including women—for the Space Shuttle program. I applied, not knowing what would happen, and was fortunate enough to be selected.

---

3. What did it mean to you to become the first American woman in space?

It was an honor, but I also understood that the mission represented progress for many women who wanted to pursue careers in science and exploration. I hoped the flight would help demonstrate that women could contribute equally to space exploration.

During training in May 1983

4. What do you remember about the moment of launch aboard the Space Shuttle Challenger?

The moment of launch aboard the Space Shuttle Challenger during the STS-7 mission was incredibly intense and exhilarating. As the countdown approached zero, the atmosphere inside the cockpit was very focused. Everyone on the crew knew exactly what their responsibilities were, and years of training had prepared us for that moment. We were strapped in our seats on the flight deck, listening carefully to the final calls from Mission Control.

When the main engines ignited, the shuttle began to shake noticeably as the engines built up thrust. A few seconds later, the solid rocket boosters ignited, and suddenly the shuttle surged upward. The acceleration was powerful, you could feel yourself pushed firmly back into the seat as the vehicle climbed away from the launch pad.

What struck me most was how quickly everything happened. Within minutes, the blue sky outside the windows gradually faded into darkness as we accelerated toward orbit. It was an extraordinary feeling to realize that we were leaving Earth and entering space.

For me, the experience also has a deeper meaning. Being part of that mission meant representing many people who had worked hard to open the doors of space exploration to a broader community. I hoped that seeing a woman fly aboard the shuttle would encourage young people—especially girls—to believe that they could pursue careers in science, engineering, and exploration.

The launch was not only the beginning of a mission; it was the beginning of a new chapter in human spaceflight.

---

5. What was your first impression when you saw Earth from space?

The first time I looked out the window and saw Earth from orbit during the STS-7 mission was an unforgettable moment. What struck me immediately was the incredible beauty of the planet. From space, Earth appears as a brilliant blue and white sphere covered with swirling clouds, deep blue oceans, and vast continents. The colors are vivid, and the curvature of the Earth is clearly visible from orbit.

Another surprising aspect was how thin the atmosphere looked. From space, you can see it as a very delicate, almost fragile layer surrounding the planet. That thin band of air is what sustains all life on Earth and seeing it from that perspective gives you a deep appreciation for how precious it really is.

The experience also changes your sense of scale. The Earth looks both vast and small at the same time. Vast in its beauty and complexity, yet small compared with the enormous expanse of space surrounding it. Perhaps the most powerful realization is that from orbit you cannot see borders or divisions between countries. You simply see one planet—our shared home. That perspective stays with astronauts long after they return to Earth. It reminds you how important it is for humanity to work together to protect the remarkable world we all share. 🌍🚀

Communicating with ground controllers from the flight deck during the STS-7 mission

---

6. During your mission, you operated on the shuttle's robotic arm. What made that task important?

Operating the shuttle's robotic arm—known as the Canadarm—was one of my primary responsibilities during the STS-7 mission aboard the Space Shuttle Challenger.

The robotic arm was a remarkable piece of technology designed to move and deploy payloads in space. During our mission, we used it to release communications satellites into orbit and to assist with experiments outside the spacecraft. The arm allowed astronauts to carefully position objects in the vacuum of space, something that would be extremely difficult to accomplish by hand.

What made the task especially important was precision. In space, even a small movement can cause large changes in position because there is very little friction or resistance. The robotic arm had to be operated slowly and carefully to ensure that satellites were deployed safely and correctly.

The Canadarm represented a new way for astronauts to work in orbit. It greatly expanded the capabilities of the Space Shuttle program, allowing crews to deploy, retrieve, and eventually repair satellites and other equipment in space. For me personally, it was exciting to operate such an advanced system. It demonstrated how human skills and robotic technology could work together to perform complex operations in the environment of space.

7. **Did your mission change the way you thought about Earth?**

Yes, it did. Seeing Earth from space emphasizes how interconnected our world is. National borders are invisible from orbit, and the planet appears as a single, shared home.

In a NASA T-38 Talon jet, Source/Author: NASA- Public Domain

8. **What challenges did you face as a woman entering a field that had traditionally been dominated by men?**

When I joined the astronaut corps at NASA in 1978, it was part of the first astronaut class to include women. Before that time, the astronaut program had been almost entirely male, largely because earlier astronauts were selected from military test pilot backgrounds where women had not historically been included.

In many ways, the environment at NASA was very professional and focused on the mission. My colleagues were primarily interested in whether you could do the job, whether you understood the spacecraft systems, could perform the technical work, and could function effectively as part of a crew.

However, there was a great deal of public attention surrounding the fact that I would become the first American woman to fly in space during the STS-7 mission aboard the Space Shuttle Challenger. Some of the questions I received from reporters at the time reflected the stereotypes that existed then—questions that male astronauts were never asked. Rather than focusing on those challenges, I tried to concentrate on the work itself. My goal was simply to do my job as well as possible and contribute to the mission.

If my flight helped open the door for other women and encouraged young girls to pursue careers in science, engineering, and exploration, then that was one of the most meaningful outcomes of the experience. Ultimately, space exploration benefits from diversity of talent and perspective. The more people who could participate, the stronger and more innovative the program becomes. 🚀🌍

9. After leaving NASA, you focused heavily on science education. Why was that so important to you?

After my missions with NASA, including the STS-7 mission aboard the Space Shuttle Challenger, I began thinking about how space exploration inspires people, especially young students.

During my time as an astronaut, I received thousands of letters from students who were fascinated by science and space. Many of them were curious about how things worked, how rockets were built, and how astronauts prepared for missions. That curiosity convinced me that education was one of the most powerful ways to shape the future.

I believe strongly that young people should have opportunities to explore science, technology, engineering, and mathematics. These fields not only help advance society but also teach students how to think critically and solve complex problems.

One area that particularly concerned me was encouraging girls to pursue careers in science and engineering. Many talented young women were interested in these subjects but sometimes lacked encouragement or role models.

By focusing on science education and outreach, I hoped to help inspire the next generation of explorers, engineers, and scientists. The future of space exploration—and many other fields—depends on the creativity and curiosity of those young minds.

In many ways, teaching and inspiring students became another kind of mission for me. Instead of exploring space directly, I was helping prepare the people who might one day travel even farther into the universe. 🚀📚🌍

---

10. What message would you share with future generations who dream of becoming astronauts?

Stay curious and pursue your interests in science and learning. Exploration requires creativity, determination, and a willingness to ask questions about the universe. Those qualities will help guide the next generation of explorers.

Floating freely on the flight deck of the Space Shuttle Challenger during the STS-7 mission. Left from her head float three Hewlett-Packard HP-41 series pocket calculators customized by NASA, which were used for various tasks on board, Source/Author: NASA – Public Domain

STS-41 G CREW INSIGNIA *--- The 41-G mission insignia focuses on its seven crew members (first to exceed six), the U.S. Flag and the Unity symbol known as the astronaut pin. The pin design in center shows a trio of trajectories merging in infinite space, capped by a bright shining star and encircled by an elliptical wreath denoting orbital flight.*

Interview with Judith Resnik

Space Shuttle Mission Specialist

Introduction

By the early 1980s, human spaceflight had entered a new era.The Space Shuttle program promised to transform space exploration from rare and experimental missions into a more routine journey into orbit. The shuttle was designed to carry scientists, engineers, satellites, and experiments into space, expanding humanity's ability to study Earth and the universe.

Among the new generation of astronauts chosen for this program was Judith Arlene Resnik. Born in Ohio, Resnik demonstrated extraordinary talent from an early age. A gifted student with a passion for science and mathematics, she pursued engineering and eventually earned a doctorate in electrical engineering. Her expertise in advanced technology and her determination to explore new frontiers made her an ideal candidate for NASA's astronaut corps.

Selected by NASA in 1978 as part of the first astronaut class to include women, Resnik quickly became known for her intelligence, discipline, and technical mastery. In 1984, she flew aboard the Space Shuttle ***Discovery*** *on mission* ***STS-41-***

***D**, becoming the **second American woman in space** after Sally Ride. During the mission, she operated the shuttle's robotic arm and helped deploy satellites, demonstrating the vital role astronauts could play in complex space operations.*

Resnik represented the future of space exploration—an era in which scientists and engineers would conduct sophisticated research in orbit. But her journey was tragically cut short.

*On January 28, 1986, she served as a mission specialist aboard the Space Shuttle **Challenger** during mission **STS-51-L**. Seventy-three seconds after launch, the shuttle was destroyed during ascent, claiming the lives of all seven astronauts on board. The disaster shocked the world.*

Yet Judith Resnik's legacy endures. Her dedication to science, exploration, and discovery continues to inspire generations of engineers, scientists, and astronauts who follow in her footsteps. She helped demonstrate that the frontier of space belongs not only to pilots, but also to the scientists and engineers who push the boundaries of knowledge.

Judith Arlene Resnik (April 5, 1949 – January 28, 1986) was an American electrical engineer, software engineer, biomedical engineer, pilot and NASA astronaut who died in the Space Shuttle *Challenger* disaster. She was the fourth woman, the second American woman and the first Jewish woman of any nationality to fly in space, logging 145 hours in orbit.

---

1. Dr. Resnik, what first inspired your interest in science and engineering?

From an early age, I was fascinated by how things work. Mathematics and science offered a way to understand complex systems, and engineering allowed me to apply those ideas to real-world challenges.

2. **How did you first learn about NASA's astronaut program?**

I learned about NASA's recruitment of scientists and engineers for the Space Shuttle program and decided to apply. The opportunity to combine engineering with space exploration was incredibly exciting.

---

3. **What did it mean to you to be selected as an astronaut?**

It was an extraordinary honor. The Space Shuttle program represented a new phase of exploration, where astronauts would conduct scientific research and operate advanced technologies in orbit.

Resnik poses with the rest of the Space Shuttle Discovery crew during the STS-41-D mission.

4. What do you remember about your first spaceflight aboard Discovery?

My first spaceflight aboard the Space Shuttle Discovery during the STS-41-D mission was an unforgettable experience. As launch approached, the atmosphere inside the shuttle was very focused. Each member of the crew had trained extensively for the mission, and we were concentrating on our specific responsibilities. Years of preparation came together in those final moments before liftoff.

When the engines ignited, the shuttle began to vibrate with tremendous energy. Then the solid rocket boosters fired, and the vehicle accelerated upward with incredible power. You could feel the force pressing you firmly into your seat as the shuttle climbed through the atmosphere.

Within only a few minutes, we reached orbit and the engines shut down. Suddenly everything became quiet, and objects inside the cabin began floating around us as we experienced weightlessness for the first time. That transition—from the powerful acceleration of launch to the calm of orbit—was remarkable. Looking out the window and seeing Earth from space was another unforgettable moment. The planet appeared incredibly beautiful and peaceful from that vantage point.

For me, the mission represented not only a technological achievement but also the opportunity to contribute to scientific research and space exploration. Being part of a shuttle crew demonstrated what dedicated teams of scientists, engineers, and astronauts could accomplish together in the pursuit of discovery. 🚀🌍

---

5. During your mission, you operated on the shuttle's robotic arm. What made that task significant?

Operating the shuttle's robotic arm, known as the Canadarm, was one of the most important responsibilities during my mission aboard the Space Shuttle Discovery on the STS-41-D mission.

The robotic arm allowed astronauts to perform extremely precise tasks outside the spacecraft without leaving the safety of the shuttle. Using it, we could deploy satellites, move equipment, and position payloads with remarkable accuracy.

What made the task especially significant was the level of precision required. In the environment of space, objects move differently than they do on Earth. Without gravity and friction, even small movements can send large structures drifting. The robotic arm had to be controlled slowly and carefully to ensure that satellites were released into the correct position and trajectory.

The Canadarm represented a major technological advance for the Space Shuttle program. It greatly expanded what astronauts could accomplish in orbit by allowing crews to handle large payloads and assist with complex operations in space.

For me personally, operating the robotic arm combined engineering skill with the excitement of spaceflight. It demonstrated how human judgment and advanced technology could work together to accomplish tasks that once seemed impossible.

In many ways, robotic arms became one of the most important tools for astronauts working in orbit, paving the way for future missions involving satellite servicing, space station construction, and scientific research in space. 🚀

---

6. What was the most memorable moment of your time in space?

There were many remarkable moments during my time in orbit aboard the Space Shuttle Discovery during the STS-41-D mission, but one experience stands out above all the others—seeing Earth from space.

Looking out the window of the shuttle, I remember being struck by the incredible beauty of our planet. The deep blue oceans, the bright white clouds swirling across the atmosphere, and the vast continents stretching across the surface created a view unlike anything I had ever seen before.

What makes that moment especially powerful is the realization that Earth appears both immense and fragile at the same time. From orbit, you can see how thin the atmosphere is and how small the planet looks compared with the vastness of space.

Another unforgettable aspect was experiencing weightlessness. Floating freely inside the spacecraft and watching objects gently drift around the cabin was something that felt almost magical at first.

But perhaps the most meaningful realization was that from space you cannot see national borders or divisions. You see only one planet—our shared home.

That perspective stays with astronauts long after they return to Earth, reminding us how important it is to protect the remarkable world we all inhabit. 🌍🚀

---

7. As one of the first women astronauts, did you feel a responsibility to inspire others?

Yes, I hoped that my work would encourage young people, especially girls, to pursue careers in science and engineering. Exploration and discovery benefit from diverse perspectives.

Resnik's flight suit on display at the Johnson Space Center
Source: Wikimedia Commons – CC0

8. What do you believe is the greatest value of space exploration?

The greatest value of space exploration lies in its ability to expand human knowledge while inspiring people across the world. When we explore space, we are not only traveling beyond Earth—we are learning more about the universe and about our own planet.

One important benefit is scientific discovery. Missions conducted by organizations such as NASA allow scientists to study the Earth from orbit, investigate other planets, and conduct experiments in microgravity that cannot be performed on the ground. These discoveries often lead to new technologies and new ways of understanding our world.

Space exploration also drives technological innovation. The challenges of designing spacecraft, satellites, and advanced instruments encourage engineers and scientists to develop new technologies that eventually benefit everyday life on Earth.

But perhaps the most powerful value is inspiration. When people see astronauts traveling into space, it reminds them that humanity is capable of extraordinary achievements. It encourages young people to pursue careers in science, engineering, medicine, and many other fields.

In the end, space exploration represents humanity's desire to understand the universe and our place within it. It reflects our curiosity, our creativity, and our willingness to reach beyond the limits of what once seemed possible. 🚀🌍✨

---

9. What advice would you give to students interested in science and engineering?

My advice to students interested in science and engineering is to stay curious and never stop asking questions. The foundation of scientific discovery begins with curiosity, the desire to understand how things work and why the world behaves the way it does.

Science and engineering can sometimes be challenging, and there will certainly be moments when problems seem difficult or complex. But those challenges are part of the learning process. Every experiment, every equation, and every project helps build the skills needed to solve real-world problems.

Resnik on the middeck of Space Shuttle Discovery during the STS-41-D mission, with an "I love Tom Selleck" sticker on her locker. A handwritten card reads: "Hi Dad".

I would also encourage students to develop strong foundations in mathematics, physics, and other scientific subjects. These disciplines provide tools that allow scientists and engineers to design new technologies, explore space, and improve life here on Earth.

Equally important is persistence. Many of the greatest discoveries in science came after repeated attempts and setbacks. The key is to remain determined and to view obstacles as opportunities to learn.

Finally, remember that science and engineering are collaborative fields. Major achievements—whether building spacecraft, developing new medical technologies, or exploring other planets—require teams of people working together. If you pursue your curiosity with dedication and enthusiasm, you may one day contribute to discoveries that help shape the future of humanity. 🚀🔬📚

---

10. What message would you share with future astronauts?

Exploration is one of humanity's greatest endeavors. Each generation of explorers builds upon the achievements of the past, expanding our understanding of the universe.

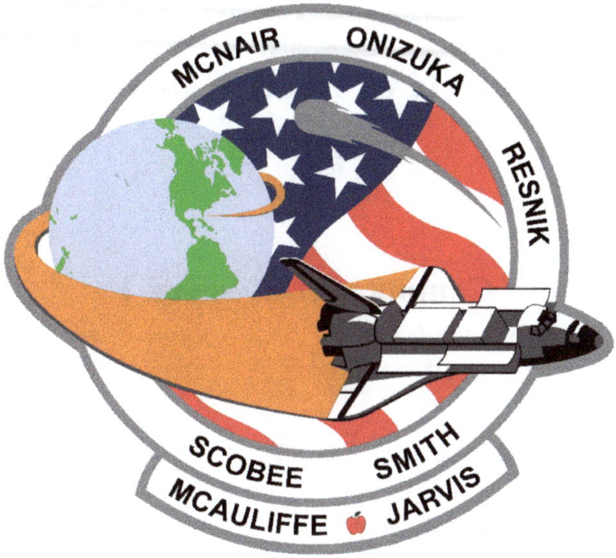

STS-51-L INSIGNIA Members of the STS-51L crew designed this patch which will represent their participation on NASA's late January 1986 mission aboard the space shuttle Challenger, depicted launching from Florida and soaring into space to carry out a variety of goals. Among the prescribed duties of the five astronauts and two payload specialists will be observation and photography of Halley's Comet, backdropped against the United States flag in the insignia. Surnames of the crew members encircle the scene, with the payload specialists being recognized below. Surname of the first teacher in space, Sharon Christa McAuliffe, is followed by a symbolic apple. Gregory Jarvis, representing Hughes, is the industrial payload specialist for the flight. NASA's crew members are astronauts Francis R. (Dick) Scobee, commander; Michael J. Smith, pilot; and Ronald E. McNair, Ellison S. Onizuka and Judith A. Resnik - all mission specialists. Public Domain.

Interview with Vladimir Komarov

Commander of Soyuz 1

Introduction

In the early years of the space race, the world watched with awe as humans began to leave Earth and venture into the vast unknown of space. Both the United States and the Soviet Union raced to achieve historic milestones, each striving to demonstrate technological leadership and scientific progress. Among the Soviet Union's most respected cosmonauts was Vladimir Mikhailovich Komarov.

Born in Moscow in 1927, Komarov grew up during a time of immense political and technological transformation. Fascinated by aviation, he pursued a career as a pilot in the Soviet Air Force, eventually becoming a skilled engineer and test pilot. His technical knowledge and disciplined flying skills earned him a place among the Soviet Union's early cosmonauts.

In 1964, Komarov flew aboard ***Voskhod 1****, becoming the first human to travel into space as part of a multi-person crew. The mission marked an important step in the* *development of more complex spaceflight operations. Komarov's professionalism and engineering expertise made him a trusted member of the Soviet space program.*

Three years later, he was selected to command a new spacecraft that represented the next stage of Soviet space technology, the ***Soyuz spacecraft****. On April 23, 1967, Komarov launched aboard* ***Soyuz 1****.*

Almost immediately after reaching orbit, serious problems began to appear. One of the spacecraft's solar panels failed to deploy, reducing the available power. Other systems malfunctioned, making it difficult to control the spacecraft and complete the planned mission. Despite these challenges, Komarov worked tirelessly to stabilize the spacecraft and prepare for a return to Earth.

After completing multiple orbits, he initiated the reentry sequence. Tragically, during the final descent, the spacecraft's main parachute failed to deploy properly. Without the parachute to slow the capsule, Soyuz 1 struck the ground at high speed. Vladimir Komarov was killed instantly. He became the first human to die during a space mission.

The tragedy shocked the Soviet space program and led to significant improvements in the design and safety of the Soyuz spacecraft. Today, the Soyuz remains one of the most reliable spacecraft ever built, carrying astronauts and cosmonauts safely to space for decades. Komarov's courage and dedication played a crucial role in the early development of human spaceflight. His sacrifice helped ensure that future missions would be safer. In the long and difficult journey toward the stars, pioneers like Vladimir Komarov helped humanity learn how to explore the universe.

Author: Valentin Cheredincev, Attribution: Mil.ru
Source: Wikimedia Commons – CC BY 4.0

Vladimir Mikhaylovich Komarov (16 March 1927 – 24 April 1967) was a Soviet test pilot, aerospace engineer, and cosmonaut. In October 1964, he commanded Voskhod 1, the first spaceflight to carry more than one crew member. He became the first Soviet cosmonaut to fly in space twice when he was selected as the solo pilot of Soyuz 1, its first crewed test flight. A parachute failure caused his Soyuz capsule to crash into the ground after re-entry on 24 April 1967, making him the first human to die in a space flight.

---

1. Colonel Komarov, what first inspired your interest in aviation?

From a young age I was fascinated by airplanes and the idea of flight. Aviation represented both adventure and technological progress, and I was eager to become part of that world.

---

2. How did you become involved in the Soviet cosmonaut program?

After serving as a pilot and studying engineering, I was selected as part of the Soviet Union's early group of cosmonauts. The program required individuals who could combine technical knowledge with flying experience.

---

3. What did it mean to you to participate in Voskhod 1, the first multi-person space mission?

Voskhod 1 was an important milestone because it demonstrated that multiple astronauts could travel together in space. It helped expand the possibilities for future missions.

Komarov with his wife Valentina Yakovlevna and daughter Irina
Creator: United Press International

---

4. What were the goals of the Soyuz program?

The Soyuz program was created as the next major step in the Soviet Union's human spaceflight efforts. After the early achievements of the Vostok and Voskhod missions, the goal was to develop a more advanced spacecraft capable of supporting longer missions and more complex operations in space.

One of the main objectives of the Soyuz program was to demonstrate orbital rendezvous and docking between spacecraft. This capability was essential for the future construction of space stations and for missions that would require spacecraft to connect and exchange crews or equipment while in orbit.

A Soyuz TMA-16 launch vehicle being transported to launchpad at Baikonur in 2009. Source: Wikimedia Commons – Public Domain

Another goal was to create a more versatile and reliable spacecraft that could support multiple crew members and remain in orbit for extended periods. The Soyuz design included separate modules for living space, reentry, and propulsion, allowing the spacecraft to operate more efficiently.

The mission of Soyuz 1 represented one of the earliest attempts to test these capabilities. Although that mission encountered serious technical difficulties, it contributed to important lessons that helped engineers improve spacecraft design.

In the long term, the Soyuz program became one of the most enduring spacecraft programs in the history of space exploration. It played a central role in later Soviet and Russian missions, including cooperation with other nations in projects such as the Apollo–Soyuz Test Project. Ultimately, the goal of Soyuz was not only to advance space technology but also to lay the foundation for humanity's continued presence in space.

Commemorative plaque and the Fallen Astronaut sculpture left on the Moon during the 1971 Apollo 15 mission.

---

5. Space exploration often involves great risks. How did astronauts and cosmonauts prepare for those dangers?

From the very beginning, everyone involved in human spaceflight understood that space exploration carried significant risks. Preparing for those dangers required extensive training, discipline, and careful engineering.

Cosmonauts in the Soviet program trained rigorously at facilities such as Star City. We spent many hours studying spacecraft systems, learning how every component of the vehicle operated. Understanding the spacecraft in detail was essential because astronauts and cosmonauts might have to solve technical problems during a mission far from Earth.

Physical preparation was also extremely important. Cosmonauts trained in centrifuges to experience the intense forces of launch and reentry, practiced survival training for emergency landings, and performed exercises designed to simulate weightlessness.

Another key element was simulation training. Crews practiced many different emergency scenarios in spacecraft simulators, learning how to respond to equipment failures, navigation problems, or communication issues. The goal was to ensure that when unexpected situations occurred, the crew would remain calm and respond methodically.

The mission of Soyuz 1 demonstrated how complex and demanding spaceflight could be. Every mission, whether successful or difficult, contributed valuable lessons that improved spacecraft design and astronaut training. Ultimately, preparation for spaceflight depended on teamwork—cosmonauts, engineers, and scientists working together to reduce risks as much as possible while continuing humanity's exploration of space.

---

6. What do you believe motivates humans to explore space?

Human beings have always been driven by curiosity and the desire to explore the unknown. Throughout history, explorers crossed oceans, climbed mountains, and traveled to distant lands to understand more about the world around them. Space exploration is simply the next step in that long tradition.

One powerful motivation is scientific discovery. By traveling beyond Earth, we learn more about the Moon, the planets, and the origins of the universe. These discoveries help humanity better understand not only space, but also our own planet and how it came to be.

Another motivation is technological progress. Programs developed by organizations such as Roscosmos and NASA have pushed the boundaries of engineering and science, leading to innovations that benefit life on Earth.

But perhaps the deepest motivation is something more fundamental—the human spirit of exploration. People naturally seek new horizons and new challenges. The idea of reaching beyond our planet and traveling into the vast universe captures the imagination of scientists, engineers, and ordinary citizens alike.

Space exploration reminds us that humanity is capable of remarkable achievements when curiosity, knowledge, and cooperation come together. In that sense, exploring space is not only about reaching other worlds—it is about fulfilling one of humanity's oldest dreams: to understand our place in the universe. 🚀🌍✨

---

7. What did it feel like to see Earth from space?

Seeing Earth from space is an experience that is difficult to describe fully with words. From orbit, our planet appears as a beautiful blue sphere surrounded by the deep blackness of space. What struck me most was the vivid contrast between the bright colors of Earth, the oceans, the clouds, and the continents—and the darkness of the universe beyond. The atmosphere forms only a very thin layer around the planet, reminding you how delicate the environment supporting life really is.

Another powerful realization is that from space you cannot see borders between countries. There are no lines separating nations or political divisions. Instead, you see one planet shared by all humanity. That perspective changes how you think about the world. It reminds you that everything we know, every culture, every city, every human story—exists on that single world.

For astronauts and cosmonauts, seeing Earth from space often creates a deep sense of appreciation and responsibility. It reinforces the idea that our planet is both magnificent and fragile, and that protecting it is one of humanity's most important tasks. 🌍🚀

---

8. How do you see the future of human spaceflight?

I believe the future of human spaceflight will unfold gradually, with humanity extending its presence farther beyond Earth step by step. The first stage will likely continue to involve long-duration missions in Earth orbit, where astronauts and cosmonauts live and work aboard space stations. These missions help scientists understand how humans can survive and operate in space for extended periods.

The next major step will be returning to the Moon and establishing permanent research bases. Such bases could serve as laboratories for studying the Moon's geology, testing new technologies, and preparing for deeper missions into the solar system.

Eventually, humanity will travel to Mars. Missions to Mars will require new propulsion systems, advanced life-support technologies, and unprecedented international cooperation. But the knowledge gained from earlier programs—both Soviet and American—will help make those journeys possible.

Space exploration has always been a collaborative effort. Organizations such as Roscosmos and NASA have shown that even during times of political tension, scientific exploration can bring nations together.

In the long run, the exploration of space will not belong to one country alone. It will become a shared endeavor of humanity, expanding knowledge, inspiring future generations, and perhaps one day allowing humans to live and work beyond Earth. 🚀🌍✨

---

9. What lesson can humanity learn from the early years of space exploration?

The early years of space exploration taught humanity one of the most important lessons in history: great achievements require courage, knowledge, and perseverance. In the 1960s, both engineers and astronauts were entering an entirely new frontier. Missions such as Soyuz 1 mission and the early American and Soviet space programs involved technologies that were still being tested and refined. Every mission carried significant risks, but each flight also brought valuable knowledge that helped improve spacecraft design and astronaut safety.

1964 USSR postage stamp honoring Vladimir Komarov - PD

Another lesson from those early years is the importance of learning from both success and failure. Every challenge encountered in space helped engineers understand how to build better systems and prepare future crews more effectively.

The early era also showed that space exploration is not only about national achievement, but also about human progress. Over time, cooperation between countries has become increasingly important, as seen in missions like the Apollo–Soyuz Test Project.

Perhaps the most enduring lesson is that exploration pushes humanity forward. When people dare to venture beyond what is known, they expand the boundaries of science, technology, and human understanding. The pioneers of spaceflight helped open the path for future generations who will travel farther into the universe. 🚀🌍✨

---

10. What message would you share with future explorers?

To the explorers who will follow us into space, my message would be this: never lose your courage or your curiosity. Space exploration demands both. The journey beyond Earth is not easy. It requires dedication, knowledge, and the willingness to face challenges that no one has encountered before. But throughout history, progress has always come from those who dared to explore the unknown.

As future astronauts travel farther into the cosmos—perhaps returning to the Moon and eventually reaching Mars—they should remember that they carry the hopes and dreams of humanity with them.

When you look back at Earth from space, you will see a small and beautiful planet without borders. That perspective reminds us that exploration is not only about discovering new worlds; it is also about understanding our shared home and our responsibility to protect it.

Work together, learn from one another, and continue pushing the boundaries of knowledge. The universe is vast, and humanity has only begun to explore it. The next great discoveries will belong to those who have the courage to take the next step into the unknown. 🚀🌍✨

Interview with Georgy Dobrovolsky

Commander of Soyuz 11

Introduction

*In June 1971, humanity reached a new frontier in space exploration. For the first time, astronauts lived and worked aboard a space station — **Salyut 1**, the world's first orbital laboratory. Leading that historic mission was Soviet cosmonaut **Georgy Dobrovolsky**, commander of **Soyuz 11**.*

*Dobrovolsky and his crewmates **Viktor Patsayev** and **Vladislav Volkov** spent **23 days in orbit**, a world record at the time, conducting experiments that helped humanity understand how people could live and work in space for extended periods. Their mission was a triumph of engineering, science, and courage. But history remembers them not only for their achievement — it remembers them for the price they paid.*

During the return to Earth, a valve malfunction caused the spacecraft cabin to depressurize. The three cosmonauts lost their lives before landing, becoming the first and only humans to die in space itself. Their sacrifice changed spacecraft design forever and ensured that future astronauts would fly safer missions.

Today, Georgy Dobrovolsky stands among the pioneers who expanded humanity's presence beyond Earth — a man whose courage carried humanity into the era of space stations. The legacy of Georgy Dobrovolsky and the crew of Soyuz 11 lives on in every space station that followed — from ***Mir*** *to today's* ***International Space Station****. Their mission opened the door to long-duration spaceflight. Their sacrifice ensured that future astronauts would fly more safely. And their courage reminds us that the path to the stars has always been written by pioneers willing to risk everything.*

Attribution: Mil.ru, Source: Wikimedia Commons – CC BY 4.0

Georgy Timofeyevich Dobrovolsky (1 June 1928 – 30 June 1971) was a Soviet cosmonaut who commanded the three-man crew of the Soyuz 11 spacecraft. They became the world's first space station crew aboard Salyut 1, but died of asphyxiation because of an accidentally opened valve. They were the first and only humans to have died in space.

---

1. Commander Dobrovolsky, when you launched aboard Soyuz 11, did you realize how historic the mission would become?

Yes, we understood that we were stepping into something entirely new. Previous missions orbited Earth, but we were going to live in space, aboard Salyut 1. It felt less like a flight and more like the beginning of humanity's permanent presence beyond Earth.

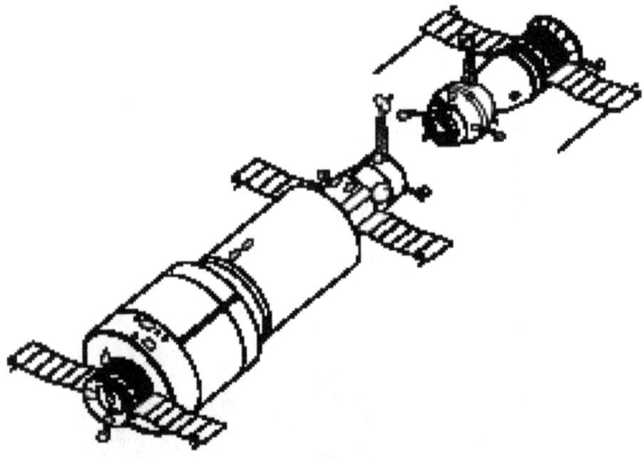

Soyuz docking with Salyut 1
Source: Wikimedia Commons – Public Domain

2. Commander Dobrovolsky, what was your first impression when you entered Salyut 1?

When we first entered the space station Salyut 1 during the mission of Soyuz 11 mission, the feeling was extraordinary. For the first time in human history, astronauts were stepping inside an orbital laboratory designed for extended work in space.

My first impression was how spacious it seemed compared with the Soyuz spacecraft. After traveling in the relatively small Soyuz capsule, entering the larger volume of the station felt almost like walking into a new world.

Inside Salyut 1 we saw rows of scientific instruments, control panels, and equipment that would allow us to conduct experiments and study the Earth from orbit. The station had been designed as a laboratory in space, and it represented a completely new stage in human spaceflight.

There was also a strong sense of responsibility. We knew we were the first crew to live and work aboard a space station, and our mission would determine whether such long-duration missions were possible.

At that moment, it felt as if humanity had taken an important step toward the future—one where astronauts could remain in orbit for long periods, conducting research and expanding our understanding of space. Standing inside

Salyut 1, we realized that space exploration was entering a new era.

---

3. During your 23 days in orbit, what moment stands out most in your memory

During our mission aboard Salyut 1 as part of the Soyuz 11 mission, many moments were unforgettable. But one of the most powerful experiences was simply looking at Earth from the windows of the station after we had settled into our work in orbit.

From that vantage point, the planet appeared incredibly beautiful and peaceful. The oceans, continents, and clouds moved slowly beneath us as the station circled Earth. Watching the sunrise and sunset every ninety minutes was especially remarkable—each transition filled the cabin with changing colors and light.

Another memorable aspect was the realization that we were living aboard the first true space station. Every experiment we conducted and every system we operated helped demonstrate that humans could remain in space for extended periods.

There was also a deep sense of responsibility. We knew that our mission would shape the future of long-duration spaceflight. The knowledge gained during those 23 days would help future astronauts live and work aboard stations for months and eventually even longer.

Looking back, the most meaningful moment was understanding that we were part of a historic step forward. Humanity was beginning to learn not just how to visit space—but how to live there. 🚀🌍

Dobrovolsky commanded the three-man crew together with Vladislav Volkov and Viktor Patsayev – Public Domain

---

4. What kind of work did you perform aboard Salyut 1?

During our mission aboard the space station Salyut 1 as part of the Soyuz 11 mission, our primary task was to demonstrate that humans could live and work in orbit for an extended period.

We carried out a wide range of scientific experiments and operational tests. One important area of work involved studying how the human body adapts to long periods in weightlessness. By monitoring our health and physical condition, scientists on the ground could learn more about the effects of spaceflight on the cardiovascular system, muscles, and balance.

Still frame from an animation of Salyut 1 with attached Soyuz. Source: Wikimedia Commons – Public Domain

We also conducted Earth observation experiments. From the station's windows and instruments, we observed weather patterns, oceans, and geological features on Earth. These observations helped scientists understand how satellites and astronauts could assist in studying our planet from orbit.

Another important part of the mission involved testing the systems of the space station itself, life-support equipment, communications systems, and onboard scientific instruments. Because Salyut 1 was the world's first space station, every system had to be carefully evaluated during our stay.

In addition, we performed astronomical observations and experiments in physics and materials science that could only be conducted in the microgravity environment of space.

In many ways, our mission served as a foundation for the future of orbital laboratories. The work we performed helped demonstrate that space stations could become permanent platforms for scientific research and human exploration beyond Earth. 🚀

Baikonur Cosmodrome's "Gagarin's Start" Soyuz launch pad
Source: Wikimedia Commons – Public Domain

---

5. Living in orbit for more than three weeks was unprecedented. What was the biggest challenge?

During the mission of Soyuz 11 mission aboard the space station Salyut 1, one of the greatest challenges was adapting both physically and mentally to living in space for such a long time.

At that time, no crew had ever remained in orbit for more than a few days. Our mission lasted **23 days**, which required us to learn how the human body responds to prolonged weightlessness. Moving, eating, sleeping, and even simple daily activities become different in microgravity.

Another challenge was maintaining a consistent routine. Life aboard a space station requires careful scheduling—scientific experiments, equipment checks, communication with mission control, and exercise to maintain physical health. Staying disciplined was essential for both the success of the mission and the well-being of the crew.

Isolation was also something we had to manage. Although we were in constant communication with engineers and scientists on the ground, we were still living in a confined environment far above Earth. Maintaining teamwork and a positive attitude among the crew was very important.

But perhaps the most significant challenge was the responsibility of being the first crew to live aboard a space station. Every system we tested and every experiment we performed would help determine whether long-duration space missions were possible.

In many ways, our mission demonstrated that humans could not only travel to space—but also live and work there for extended periods, opening the door to the future of orbital stations and long-term exploration. 🚀

---

6. How important was teamwork among the Soyuz 11 crew?

Teamwork was essential during the mission of Soyuz 11 mission aboard the space station Salyut 1. In space, astronauts and cosmonauts rely on one another completely. Every task—from operating spacecraft systems to conducting scientific experiments—requires coordination and mutual trust. Each member of the crew must understand not only their own responsibilities but also how their work supports the others.

Our crew included Vladislav Volkov and Viktor Patsayev. We spent many months training together before the mission, practicing procedures again so that we could function as a single, well-coordinated team.

Living and working in the confined environment of a space station for more than three weeks also required patience and understanding. Cooperation was essential not only for completing the mission tasks but also for maintaining morale and a positive atmosphere aboard the station.

Space exploration has always depended on teamwork—not only among the crew in orbit but also with the engineers, scientists, and mission controllers on the ground. The success of any mission reflects the combined effort of many people working together toward a shared goal. In that sense, teamwork is one of the most important qualities for anyone who ventures into space. 🚀

---

7. Did you ever feel fear during the mission?

Every cosmonaut understands the risks. But fear cannot control you. Our duty was to explore, to learn, and to push humanity forward. Courage does not mean the absence of fear — it means acting despite it.

---

8. What did it feel like to orbit Earth for so long?

Orbiting Earth for more than three weeks during the mission of Soyuz 11 mission aboard Salyut 1 was an extraordinary experience. At first, the sensation of living in microgravity was remarkable. Without gravity, every movement becomes different—objects floating, and you must learn new ways to move, work, and even sleep. It takes time for the body to adjust, but eventually it becomes a natural way of living.

One of the most fascinating aspects of orbiting Earth was the constant view of the planet below us. Every ninety minutes we completed another orbit, which meant we witnessed many sunrises and sunsets each day. Watching the light spread across the horizon and illuminate the atmosphere was always breathtaking.

But living in orbit for such a long time also requires discipline. Our days were carefully scheduled with experiments, equipment checks, communication with mission control, and physical exercise to maintain our health.

Perhaps the most powerful feeling was the realization that we were among the first humans to truly live in space, not just visit it briefly. That experience suggested a future in which humans might remain in orbit for months or even years, conducting research and preparing for journeys farther into the solar system. In many ways, those weeks in orbit felt like a glimpse of humanity's future beyond Earth. 🚀🌍

---

9. What do you believe the future of space stations will be like?

The mission of Soyuz 11 mission aboard Salyut 1 represented the beginning of a new era in human spaceflight. It demonstrated that humans could live and work in orbit for extended periods of time.

In the future, I believe space stations will become permanent laboratories in space, where astronauts and scientists from many nations work together. These stations will allow researchers to conduct experiments in microgravity, observe Earth, and study the universe in ways that are impossible from the ground.

Later stations—such as Mir and the International Space Station—have already expanded on the ideas first tested with Salyut 1. They show that long-duration missions and international cooperation are possible.

In the future, space stations may also serve as staging points for deeper exploration. Astronauts traveling to the Moon, Mars, or beyond could assemble spacecraft, conduct training, and prepare for long missions while in orbit.

Ultimately, space stations may become small communities in space—places where humans learn how to live beyond Earth and build the foundation for future exploration of the solar system. What began with the first stations like Salyut may one day lead to a permanent human presence beyond our planet. 🚀🌍✨

---

10. If humanity remembers one thing about Soyuz 11, what should it be?

If there is one thing humanity should remember about the mission of Soyuz 11 mission, it is that it marked the beginning of humanity learning how to live and work in space.

Our crew—together with Vladislav Volkov and Viktor Patsayev—became the first humans to occupy a space station, Salyut 1. During our 23 days in orbit, we demonstrated that long-duration missions aboard an orbital laboratory were possible.

Although the mission ended tragically during the return to Earth, the lessons learned from Soyuz 11 greatly improved the safety of future spacecraft and helped shape the design of later missions.

Every space station that followed—from Salyut and Mir to the International Space Station—built upon the foundation established during that mission.

For that reason, Soyuz 11 should be remembered not only for its sacrifice, but also for its contribution to the future of human spaceflight. It showed that humanity could begin to move beyond brief visits to space and start building a sustained presence in orbit.

In the history of exploration, progress often comes with great challenges. The legacy of Soyuz 11 is that it helped open the door to the era of space stations and long-duration missions that continue today. 🚀🌍✨

Soyuz program, Source: Wikimedia Commons – Public Domain

Interview with Victor Patsayev

Cosmonaut of Soyuz 11

Introduction

Among the early pioneers of human spaceflight, ***Viktor Patsayev*** *occupies a unique place in history. Born in Kazakhstan in 1933, Patsayev was not only a cosmonaut but also a scientist and engineer, deeply fascinated by astronomy and space technology. His passion for exploring the universe eventually led him to join the Soviet cosmonaut corps and take part in one of the most historic missions of the early space age.*

In June 1971, Patsayev launched aboard ***Soyuz 11*** *together with commander* ***Georgy Dobrovolsky*** *and flight engineer* ***Vladislav Volkov****. Their destination was* ***Salyut 1****, the world's first orbital space station. During their* ***23-day mission****, the crew conducted scientific experiments, studied the Earth and the stars, and demonstrated for the first time that humans could live and work aboard an orbital station for extended periods.*

Patsayev made history in another remarkable way: he became the first human to operate a telescope in space, performing astronomical observations from orbit — a glimpse of the future where space stations would become platforms for studying the universe. But the triumph of the mission ended in tragedy. During the return to Earth, the Soyuz spacecraft depressurized during re-entry, and all three cosmonauts lost their lives.

Today, Viktor Patsayev is remembered not only as a brave explorer but also as the first astronomer to observe the cosmos from space, a pioneer who helped transform space stations into laboratories for science.

The mission of ***Soyuz 11*** *marked the beginning of long-duration human presence in space. The work of Viktor Patsayev helped transform space stations into scientific observatories — a concept that continues today aboard the* ***International Space Station****. His legacy lives on every time astronomers observe the universe from space and every time scientists conduct research aboard orbiting laboratories.*

Photographed by Alexander Mokletsov – Public Domain

Viktor Ivanovich Patsayev (19 June 1933 – 30 June 1971) was a Soviet cosmonaut who flew on the Soyuz 11 mission and was part of the third space crew to die during a space flight. On board the space station Salyut 1 he operated the Orion 1 Space Observatory (see Orion 1 and Orion 2 Space Observatories); he became the first man to operate a telescope outside the Earth's atmosphere

---

1. Viktor, what first inspired your fascination with space and astronomy?

Since childhood I was fascinated by the stars. I often looked at the night sky wondering what lay beyond it. When humanity began launching satellites and sending people into space, I knew that exploring the universe was no longer just a dream — it had become our future.

---

2. How did your scientific background shape your role as a cosmonaut?

Many people imagine cosmonauts as pilots, but we were also scientists. My engineering and astronomical training helped me conduct experiments aboard the station and operate instruments designed to study the Sun and distant stars.

---

3. What was it like arriving at Salyut 1, the world's first space station?

It felt as though we had entered a laboratory floating above Earth. For the first time, we were not just visitors to space — we were residents of an orbital research station.

---

4. Cosmonaut Patsayev, you became the first human to operate a telescope in space. What did that experience feel like?

Using a telescope in space during the mission of Soyuz 11 mission aboard Salyut 1 was an extraordinary experience. The telescope we used, known as the Orion telescope, allowed us to observe stars and other celestial objects without the distortion caused by Earth's atmosphere. From space, the view of the universe is much clearer and more stable, making it possible to collect valuable astronomical data.

When I looked through the instrument and saw the stars from orbit, I felt a deep sense of wonder. The stars appeared incredibly sharp and bright against the black background of space. It was as if we were seeing the universe in its purest form.

What made the moment especially meaningful was the realization that we were conducting astronomical observations from space for the first time. For centuries, astronomers had studied the universe from the ground, limited by the atmosphere and weather. Now, from orbit, we could observe the cosmos with new clarity.

The experience reminded me that space stations are not only platforms for living in space—they are also laboratories for exploring the universe itself. In that moment, I felt that we were opening a new chapter in astronomy, one that would eventually lead to even more powerful space telescopes and discoveries about the universe. 🔭🚀✨

---

5. What scientific work did you perform aboard Salyut 1?

During the mission of Soyuz 11 mission aboard Salyut 1, our crew conducted a variety of scientific experiments designed to take advantage of the unique conditions of space.

One of my main responsibilities involved astronomical observations using the Orion space telescope. This instrument allowed us to study stars and ultraviolet radiation from space, free from the interference of Earth's atmosphere. It was the first time a human had operated a telescope while in orbit, which made the observations particularly meaningful.

We also conducted Earth observation experiments, studying weather systems, oceans, and geographic features from orbit. These observations helped scientists understand how space-based platforms could be used to monitor our planet.

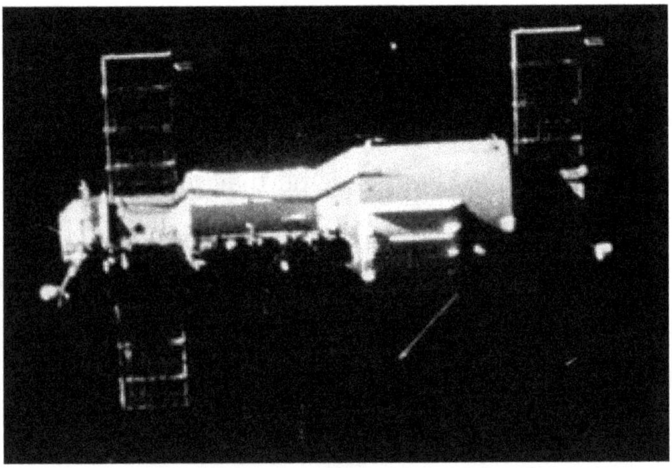

Salyut 1 as seen from the departing Soyuz 11
Source: Wikimedia Commons – Public Domain

Another important area of research involves human physiology in microgravity. Since we were spending more than three weeks in orbit, doctors and scientists were very interested in learning how the human body adapts to long-duration spaceflight.

In addition, we performed experiments in materials science and physics, studying how certain processes behave in weightlessness.

Because Salyut 1 was the world's first space station, every experiment we conducted helped establish the foundations for future orbital laboratories. Our work demonstrated that space stations could serve as important scientific platforms for research that cannot be performed on Earth. 🚀🔭🌍

---

6. What was daily life like aboard the station?

Daily life aboard Salyut 1 during the Soyuz 11 mission was both fascinating and demanding. Although we were orbiting Earth far above the planet, our days followed a carefully planned routine. Each day began with communication with mission control on the ground. Engineers and scientists would review our schedule, which included scientific experiments, maintenance tasks, and system checks for the station.

Much of our time was devoted to scientific research. We conducted astronomical observations, monitored Earth's weather and geography, and performed experiments related to physics and the effects of microgravity on the human body.

Living in microgravity requires adjusting to new ways of performing even simple activities. Moving around the station meant gently pushing off walls or handrails rather than walking. Eating and drinking were also different, since food and liquids had to be carefully contained so they would not float away.

Exercise was important as well. Spending long periods in weightlessness can weaken muscles, so we followed physical routines to help maintain our health. Despite the busy schedule, there were moments when we could look out the station's windows and watch Earth pass below us. Every ninety minutes the station completed another orbit, giving us frequent views of continents, oceans, and the changing colors of sunrise and sunset.

Life aboard the station required discipline, teamwork, and adaptation—but it also offered a remarkable perspective on our planet and the universe beyond. 🚀🌍

---

7. What was the most surprising thing about living in space?

One of the most surprising aspects of living aboard Salyut 1 during the Soyuz 11 mission was how quickly the human body begins to adapt to weightlessness. At first, everything feels unfamiliar. Without gravity, you cannot walk in the usual way. Instead, you move by gently pushing yourself from one surface to another and floating through the station. Even simple tasks such as eating, writing, or using tools require learning new techniques.

Another surprising experience was how ordinary life eventually began to feel. After several days in orbit, floating became natural, and the station started to feel like a normal workplace—just one that happened to be traveling around Earth at incredible speed.

Perhaps the most remarkable surprise, however, was the constant view of our planet. Every ninety minutes we completed another orbit, which meant we saw many sunrises and sunsets each day. Watching the thin atmosphere glowing along the horizon was always breathtaking.

Living in space changes your perspective. From orbit, you see Earth as one beautiful world without borders, surrounded by the vast darkness of space. That realization is something many astronauts and cosmonauts never forget. 🌍🚀

---

8. Did the view of Earth from orbit change your perspective?

Absolutely. When you see Earth from space, you realize how fragile and beautiful it is. There are no borders visible — only a small world shared by all humanity.

---

9. What do you believe space stations will become in the future?

The mission of Soyuz 11 mission aboard Salyut 1 showed that humans could live and work in orbit for extended periods of time. In many ways, that mission was only the beginning of what space stations could become.

In the future, I believe space stations will develop into large international laboratories in orbit. Scientists from many countries will work together there to conduct experiments in physics, biology, medicine, and astronomy that cannot be performed on Earth.

Later stations such as Mir and the International Space Station have already demonstrated the value of long-term cooperation in space. Space stations may also serve as gateways for deeper exploration. Astronauts traveling to the Moon, Mars, or other destinations could assemble spacecraft, test technologies, and prepare for long missions while in orbit.

In the distant future, stations may even grow large enough to support small communities of scientists, engineers, and explorers living in space for long periods. What began with the first orbital laboratory may eventually become humanity's first permanent foothold beyond Earth. Space stations will likely play a central role in humanity's journey deeper into the universe. 🚀🌍✨

---

10. If future generations remember your mission, what would you hope they remember?

That our mission proved something important: space is not only a place for exploration, but also a place for discovery and science. If future astronomers observe the universe from orbit, I hope they remember that we helped begin that journey.

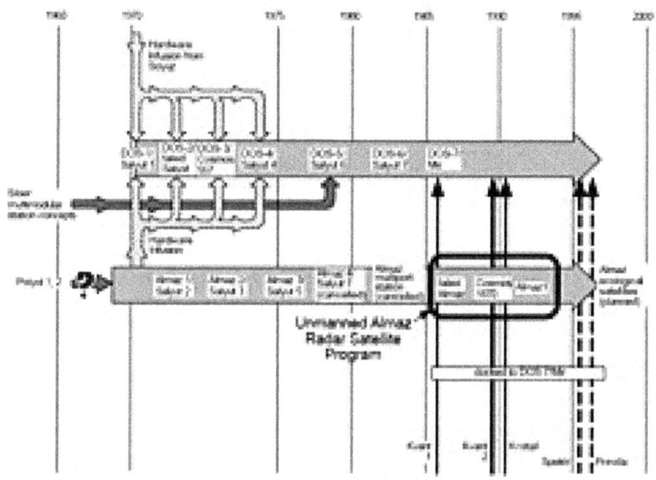

Development of Soviet space stations:

- *The large horizontal arrows trace the evolution of the two Soviet space station programmes DOS (top) and Almaz-OPS (bottom)*
- *Dark gray arrows trace the infusions from the Soyuz and OPS programmes to DOS*
- *Solid and dashed black arrows indicate modules intended for Mir, containing influences from OPS with the addition of space tugs*

Source: Wikimedia Commons – Public Domain

Interview with Vladislav Volkov

Flight Engineer of Soyuz 11

Introduction

Few pioneers of space exploration experienced both the triumph and the tragedy of the early space age as profoundly as Vladislav Volkov.

Born in Moscow in 1935, Volkov began his career as an aerospace engineer, working on spacecraft development during the formative years of the Soviet space program. His technical brilliance and calm determination eventually led him to become a cosmonaut — one of the select few chosen to carry humanity beyond Earth.

Volkov first flew in space aboard Soyuz 7, participating in one of the earliest multi-spacecraft missions designed to test coordinated operations in orbit. That experience prepared him for an even greater challenge.

In June 1971, Volkov launched again — this time aboard Soyuz 11, alongside commander Georgy Dobrovolsky and engineer Viktor Patsayev. Their destination was Salyut 1, the world's first orbital space station.

*During their historic **23-day mission**, the crew demonstrated that humans could live and work in space for extended periods. They conducted experiments, tested life-support systems, and opened the door to the era of long-duration spaceflight.*

But their mission ended in tragedy during the return to Earth when a depressurization accident occurred in the Soyuz capsule. All three cosmonauts lost their lives, becoming the only humans in history to die in the vacuum of space.

*Today, Vladislav Volkov is remembered as both an **engineer and explorer**, a man who helped lay the technological foundation for the space stations that followed.*

Photographed by Valentin Cheredintsev, Source Wikimedia Commons – Public Domain

Vladislav Nikolayevich Volkov (23 November 1935 – 30 June 1971) was a Soviet cosmonaut who flew on the Soyuz 7 and Soyuz 11 missions. The second mission terminated fatally. Volkov and the two other crew members were asphyxiated on reentry, the only three people to have died in outer space.

---

1. Vladislav, you began your career as an aerospace engineer. What inspired you to become a cosmonaut?

Working as an engineer on spacecraft made me realize that designing machines to explore space was only part of the adventure. I wanted to experience that journey myself — to see the Earth from orbit and help expand human knowledge beyond our planet.

---

2. You flew first on Soyuz 7. How did that mission prepare you for Soyuz 11?

Soyuz 7 taught us the complexity of operating spacecraft in orbit and coordinating with other missions. It was an important step toward more ambitious operations — such as docking with space stations.

---

3. What was your reaction when you first entered Salyut 1?

It was remarkable. For the first time we had a place in space where we could move freely and conduct experiments for weeks. It felt like the beginning of a new era.

---

4. As an engineer, what technical challenges interest you most aboard the station?

As an engineer, one of the most fascinating aspects of our mission aboard Salyut 1 during the Soyuz 11 mission was the opportunity to evaluate how a complex space laboratory operates over an extended period.

The station contained many interconnected systems—life-support equipment, electrical power systems, communications instruments, and scientific devices. One of our most important responsibilities was to monitor how these systems performed during long-duration operations in orbit.

Life-support systems were especially important. Maintaining a stable environment inside the station requires careful control of air pressure, oxygen levels, temperature, and humidity. Studying how these systems functioned helped engineers improve the design of future space stations.

Another technical challenge involved communications and control systems. Because the station was constantly moving around Earth at high speed, maintaining reliable communication with mission control required sophisticated equipment and coordination.

We were also interested in the performance of scientific instruments in microgravity. Operating equipment in weightlessness can produce very different results compared with experiments conducted on Earth, so part of our role was to evaluate how these instruments behaved in orbit.

For an engineer, the station was essentially a new type of laboratory—one that allowed us to test technologies that would eventually make longer missions and more advanced space stations possible. Every system we studied aboard Salyut 1 helped engineers understand how humans could live and work in space for longer periods in the future. 🚀

---

5. What was daily life like aboard the first space station?

Every day was filled with experiments, maintenance tasks, and communication with mission control. But we also had moments to simply observe Earth. Those quiet moments reminded us why exploration matters.

---

6. Did you feel that you were opening the door for future space stations?

Yes, very much so. During the mission of Soyuz 11 mission aboard Salyut 1, we understood that we were participating in something entirely new. Before that time, most space missions had lasted only a few days. Astronauts and cosmonauts would travel into orbit, conduct experiments, and then return to Earth. With Salyut 1, however, the goal was different: to demonstrate that humans could live and work in space for extended periods.

Every task we performed—monitoring the station's systems, conducting experiments, and adapting to daily life in microgravity—helped engineers and scientists understand how future stations should be designed.

In that sense, our mission was not only about the present. It was about the future of human spaceflight. Later space stations such as Mir and the International Space Station continued the work that began with Salyut 1. They showed that astronauts could remain in orbit for months and even longer.

We believed that our mission was opening the first door to a new era—one in which space stations would become permanent laboratories in orbit and perhaps one day even small communities beyond Earth. What we began aboard Salyut 1 was only the first step in humanity learning how to live in space. 🚀🌍✨

A painting of Apollo–Soyuz, a first milestone in international spaceflight, Source: Wikimedia Commons – Public Domain

---

7. What surprised you most about living in space?

How adaptable humans are. After a few days, floating becomes natural, and even complex tasks can be performed efficiently.

---

8. What do you feel when looking at Earth from orbit?

It is difficult to describe. Earth appears peaceful and incredibly beautiful. You realize how small our world is compared to the vastness of space.

---

9. What do you think will be the next major milestones in human spaceflight?

I believe humanity will build larger orbital stations, establish permanent crews in space, and eventually travel to the Moon and Mars.

---

10. If history remembers your mission, what would you want people to understand about it?

That exploration requires courage and perseverance. Our mission proved that humans can live and work in space. That knowledge will guide future generations toward even greater discoveries.

A Commemorative Plaque honoring Space Station Intergovernmental Agreement signed on 28 January 19

Interview with Christa McAuliffe

The Teacher in Space

Introduction

In the mid-1980s, NASA sought to bring space exploration closer to everyday life. The Space Shuttle program had already carried scientists, engineers, and astronauts into orbit, but NASA believed the next step was to involve ordinary citizens in the journey. The idea was simple but powerful: to send a teacher into space.

The ***Teacher in Space Project*** *was created to inspire students across the United States and to demonstrate that space exploration belonged not only to astronauts and scientists, but to everyone. From more than 11,000 applicants, NASA selected a high school social studies teacher from New Hampshire: Christa McAuliffe.*

Born Sharon Christa Corrigan, McAuliffe had always been fascinated by history and exploration. As a teacher, she loved helping students understand the great adventures that shaped human civilization—from early explorers crossing oceans to the pioneers who first traveled into space.

When she was chosen for the mission, McAuliffe embraced the opportunity with enthusiasm and humility. She planned to teach lessons from orbit, sharing the experience of spaceflight directly with classrooms around the world. Millions of students eagerly anticipated the mission.

On January 28, 1986, McAuliffe launched aboard the Space Shuttle ***Challenger*** *as part of mission* ***STS-51-L****, alongside six professional astronauts. The shuttle rose into the clear morning sky from Cape Canaveral as spectators and schoolchildren across the nation watched.*

Seventy-three seconds after launch, tragedy struck. The Challenger shuttle broke apart during ascent, claiming the lives of all seven crew members. The disaster shocked the world and deeply affected the global space community. Yet Christa McAuliffe's legacy endured.

Her dream of bringing education and exploration together inspired countless students to pursue careers in science, engineering, and teaching. Schools, scholarships, and educational programs have since been named in her honor, continuing her mission of inspiring young minds. Christa McAuliffe believed that exploration was part of the human story—and that education was the key to understanding it. Her journey reminds us that the desire to learn and explore is one of humanity's most powerful forces.

Sharon Christa McAuliffe (née **Corrigan**; September 2, 1948 – January 28, 1986) was an American teacher and astronaut from Concord, New Hampshire, who died on the Space Shuttle *Challenger* on mission STS-51-L, where she was serving as a payload specialist.

---

1. Mrs. McAuliffe, what inspired you to become a teacher?

I have always believed that education is one of the most powerful ways to inspire curiosity and creativity. Teaching allows you to help students see connections between the past, the present, and the future.

2. How did you react when you first heard about NASA's Teacher in Space program?

I was immediately excited. The idea of bringing space exploration directly into the classroom was extraordinary. I felt it would be a wonderful opportunity to help students experience history as it was being made.

---

3. What did it mean to you to be selected from thousands of applicants?

It was both thrilling and humbling. I knew that many talented teachers had applied, and I felt a responsibility to represent educators and students across the country.

---

4. What lessons did you plan to teach from space?

I planned to share lessons about life aboard the Space Shuttle and about the scientific experiments conducted in orbit. My goal was to help students feel connected to the mission and to see space exploration as something they could be part of.

McAuliffe undergoing pre-flight training
experiences weightlessness during a KC-135 "vomit comet" flight
Author/Source: NASA: Public Domain

---

5. What excited you most about the opportunity to travel into space?

The chance to experience space firsthand and then share that experience with students around the world was incredibly exciting. It was a unique opportunity to combine education and exploration.

---

6. How did you prepare for the challenges of astronaut training?

The training was rigorous but fascinating. Learning about spacecraft systems, microgravity, and mission procedures gave me a new appreciation for the complexity of spaceflight.

---

7. What message did you hope your mission would send to students?

I hoped students would see that exploration and discovery are part of everyday life. Science and learning are not limited to laboratories; they are part of the adventure of understanding the world.

---

8. Why do you think space exploration inspires so many young people?

Space represents the ultimate frontier. It challenges us to imagine new possibilities and to push the boundaries of what we can achieve.

Challenger crew from left to right: (front row) Michael J. Smith, Dick Scobee, Ronald McNair; (back row) Ellison Onizuka, Christa McAuliffe, Gregory Jarvis, and Judith Resnik

9. **What advice would you give to students who dream of exploring space?**

Stay curious and continue learning. The future of exploration will depend on the creativity and determination of young people who are willing to pursue their dreams.

---

10. **What message would you share with teachers around the world?**

Teachers have the extraordinary privilege of shaping the future. By encouraging curiosity and imagination, they help students see that they can play a role in humanity's greatest adventures.

McAuliffe and Barbara Morgan

Interview with Ronald McNair

Physicist and Space Shuttle Astronaut

Introduction

The Space Shuttle era opened a new chapter in human space exploration—one in which scientists, engineers, and researchers joined astronauts in orbit to conduct experiments that would expand humanity's understanding of the universe. Among the remarkable individuals chosen to participate in this new generation of missions was Ronald Erwin McNair.

Born in 1950 in Lake City, South Carolina, McNair grew up during a time when opportunities for African Americans in science and education were limited. Yet from an early age, he demonstrated extraordinary determination and intellectual curiosity.

His passion for science led him to pursue physics, and through dedication and hard work he earned a doctorate from the ***Massachusetts Institute of Technology (MIT)****. McNair became a respected physicist specializing in laser technology—an emerging field with applications ranging from scientific research to advanced communications.*

But McNair's talents extended beyond science. He was also an accomplished musician who loved playing the saxophone. Friends and colleagues often described him as a man who combined scientific brilliance with creativity and warmth.

In 1978, NASA selected McNair as part of the same astronaut class that included pioneers such as Sally Ride and Judith Resnik. This class marked a turning point in NASA's history, opening the astronaut corps to scientists, women, and individuals from diverse backgrounds. McNair flew his first mission aboard the Space Shuttle ***Challenger*** *in 1984, where he conducted experiments and helped deploy satellites in orbit. His work contributed to the growing role of scientific research aboard the Space Shuttle.*

Two years later, he returned to Challenger as a mission specialist for ***STS-51-L****. On January 28, 1986, the shuttle launched from Cape Canaveral. Seventy-three seconds into the flight, the spacecraft broke apart during ascent. All seven crew members were lost. The tragedy shocked the world and led to a major reassessment of spaceflight safety.*

Yet Ronald McNair's legacy continues to inspire future generations. Scholarships, schools, and scientific programs bear his name, encouraging students to pursue careers in science, engineering, and exploration. His life demonstrated that curiosity, determination, and imagination can carry individuals far beyond the limits once imposed by circumstance. Ronald McNair believed deeply in the power of knowledge—and in humanity's ability to explore the universe through science.

Ronald Erwin McNair (October 21, 1950 – January 28, 1986) was an American NASA astronaut and physicist. He died at the age of 35 during the launch of the Space Shuttle *Challenger* on mission STS-51-L, in which he was serving as one of three mission specialists in a crew of seven. Prior to the *Challenger* disaster, McNair flew as a mission specialist on STS-41-B aboard *Challenger* from February 3 to 11, 1984, becoming the second African-American in space.

---

1. Dr. McNair, what first inspired your interest in science?

From an early age, I was fascinated by how the universe works. Physics offered a way to explore the fundamental laws that govern nature. I was especially interested in how light and energy behave.

---

2. How did your academic journey lead you to NASA?

After studying physics in college, I pursued graduate research in laser physics at MIT. That work eventually led me to apply for NASA's astronaut program when they began recruiting scientists for the Space Shuttle missions.

---

3. What did it mean to you to be selected as an astronaut?

It was an incredible opportunity. The Space Shuttle program allowed scientists to conduct experiments in the unique environment of space, expanding our ability to study physical phenomena.

---

4. What do you remember about your first launch aboard the Space Shuttle?

The launch experience is unforgettable. The engines ignite with tremendous power, and within minutes you are traveling faster than any aircraft on Earth.

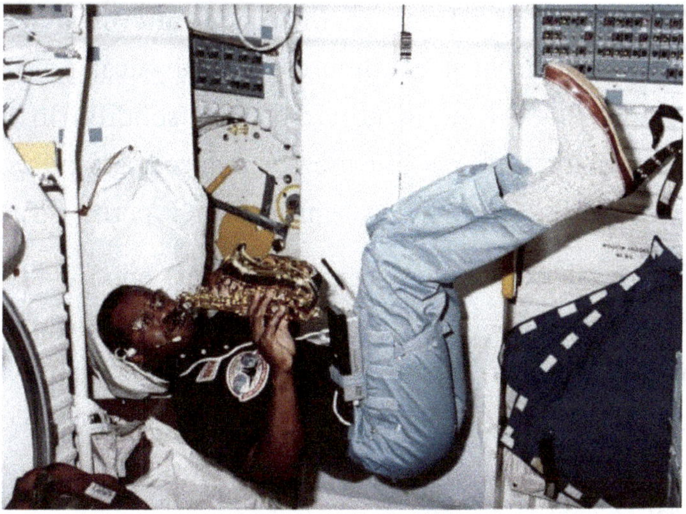

McNair plays the saxophone aboard Space Shuttle Mission 41-B. Source: Wikimedia Commons Public Domain

---

5. What was your first impression when you saw Earth from orbit?

The beauty of Earth from space is extraordinary. The oceans, clouds, and continents form a breathtaking view that reminds you how special our planet is.

---

6. Dr. McNair, as a physicist, what scientific opportunities does spaceflight provide?

Spaceflight provides scientists with an extraordinary laboratory where the laws of physics can be studied under conditions that are impossible to reproduce on Earth. One of the most important advantages of space is **microgravity**. On Earth, gravity influences almost every physical process—from how fluids flow to how materials solidify. In orbit, where gravity's effects are greatly reduced, scientists can observe how these processes behave in their purest forms.

For physicists, this environment creates opportunities to study **fluid dynamics, crystal growth, combustion, and materials science** in new ways. These experiments often reveal fundamental principles that help improve technologies used on Earth, from advanced manufacturing techniques to improved energy systems.

Spaceflight also enables the development and testing of **highly sensitive scientific instruments**. Observatories and detectors placed in space can measure cosmic radiation, study distant stars and galaxies, and explore the physical structure of the universe without interference from Earth's atmosphere.

Another important opportunity is the ability to combine different scientific disciplines. Space missions often involve collaboration among physicists, engineers, biologists, and medical researchers, creating a truly interdisciplinary environment for discovery. In many ways, space is not just a destination, it is a powerful laboratory that allows humanity to deepen its understanding of the fundamental laws that govern nature. 🚀🔬🌌

---

7. You were also a musician who loved playing the saxophone. Did music play an important role in your life?

Absolutely. Music and science both require creativity and discipline. Playing the saxophone allowed me to express another side of my personality beyond my work in physics.

---

8. What do you believe is the greatest value of space exploration?

Space exploration expands knowledge and inspires people around the world. It encourages scientific progress and helps humanity better understand our place in the universe.

---

9. What message would you share with students interested in science?

Believe in your abilities and pursue your curiosity. Scientific discovery begins with the desire to ask questions and seek answers.

---

10. What message would you share with future astronauts?

Exploration requires courage, preparation, and teamwork. The discoveries awaiting humanity in space will continue to expand our understanding of the cosmos.

Dr. Ronald E. McNair memorial in his hometown, Lake City, South Carolina – Public Domain

Interview with Kalpana Chawla

Space Shuttle Columbia Astronaut

Introduction

In every generation, there are individuals whose dreams carry them far beyond the boundaries of their childhood surroundings. Kalpana Chawla was one of those dreamers.

Born in Karnal, a small city in northern India, Chawla grew up fascinated by airplanes and the vast sky above her home. As a young girl, she would often watch aircraft flying overhead and imagine what it would be like to explore the heavens beyond Earth.

At a time when few women pursued careers in aerospace engineering, Chawla followed her passion with determination. She studied aeronautical engineering in India before moving to the United States to continue her education. There she earned advanced degrees in aerospace engineering and eventually became a naturalized American citizen.

Her academic achievements and expertise in aerodynamics led her to NASA, where she contributed to research on complex aircraft and spacecraft systems. In 1994, NASA selected Kalpana Chawla as an astronaut.

Three years later, in 1997, she flew aboard the Space Shuttle ***Columbia*** *on mission* ***STS-87****, becoming the* ***first woman of Indian origin to travel into space****. Her journey inspired millions of people around the world, particularly students who saw in her story the power of determination and education.*

In orbit, Chawla helped conduct scientific experiments that studied the behavior of materials and physical processes in microgravity—research that could not be performed on Earth. Her calm professionalism and deep enthusiasm for science made her a respected member of the astronaut corps.

In 2003, she returned to space aboard ***Columbia*** *for mission* ***STS-107****, a flight dedicated entirely to scientific research in orbit. During the sixteen-day mission, the crew conducted dozens of experiments in biology, physics, and materials science. But tragedy struck during the spacecraft's return to Earth.*

On February 1, 2003, as Columbia reentered Earth's atmosphere, damage sustained during launch caused the shuttle to break apart. All seven astronauts aboard the spacecraft were lost.

The world mourned the loss. Yet Kalpana Chawla's legacy continues to inspire generations of students and explorers. Schools, scholarships, and scientific institutions around the world honor her name. Her journey—from a small town in India to the frontier of space—remains a powerful reminder that curiosity, determination, and education can carry human dreams far beyond Earth.

Kalpana Chawla (March 17, 1962 – February 1, 2003) was an American astronaut and aerospace engineer who was the first woman of Indian origin to fly to space. Chawla expressed an interest in aerospace engineering from an early age and took engineering classes at Dayal Singh College and Punjab Engineering College in India. She then traveled to the United States, where she earned her MSc and PhD, becoming a naturalized United States citizen in the early 1990s.

---

1. Dr. Chawla, when did your fascination with aviation and space begin?

My fascination began during childhood. I often watched airplanes flying above my hometown and wondered what it would feel like to fly among the clouds and beyond them. That curiosity led me toward studying aeronautical engineering.

---

2. What challenges did you face pursuing aerospace engineering as a young student?

At the time, relatively few women pursued careers in aerospace engineering. However, I was fortunate to have supportive teachers and family members who encouraged me to follow my interests.

---

3. What motivated you to move to the United States to continue your studies?

The United States offered excellent opportunities to study aerospace engineering and participate in advanced research. I wanted to deepen my understanding of flight and eventually contribute to the exploration of space.

The crew of STS-87 in September 1997. From left to right - in orange: Chawla, Lindsey, Kregel, Kadeniuk; in white: Scott, Doi

4. Dr. Chawla, how did it feel when NASA selected you as an astronaut?

When I learned that I had been selected as an astronaut by NASA, it was one of the most exciting and humbling moments of my life. From a very young age, I had been fascinated by airplanes, the sky, and the possibility of traveling into space. Growing up in India, I often looked up at the stars and imagined what it would be like to explore beyond Earth. Becoming an astronaut felt like the realization of a dream that had guided me for many years.

At the same time, the selection also brought a strong sense of responsibility. Being chosen meant joining a team of highly skilled scientists, engineers, and explorers who were dedicated to advancing human knowledge through space exploration.

Astronaut training was demanding and required learning many new skills—from operating spacecraft systems to working in microgravity and conducting scientific experiments. But every part of the process was inspiring, because it represented an opportunity to contribute to humanity's exploration of space.

For me, the moment of selection was not just a personal achievement. It was also a reminder that dreams can cross borders and cultures. Space exploration belongs to all humanity, and anyone with curiosity, dedication, and perseverance can become part of that journey. 🚀🌍✨

---

5. What do you remember about your first launch aboard the Space Shuttle?

My first launch aboard the Space Shuttle Columbia during the STS-87 mission was an unforgettable moment. As the countdown approached its final seconds, the atmosphere inside the shuttle was calm but intensely focused. Each member of the crew was concentrating on their responsibilities, reviewing procedures and listening carefully to the final calls from Mission Control.

When the main engines ignited, the shuttle began to vibrate as the enormous power of the engines built up beneath us. A few seconds later, the solid rocket boosters ignited, and suddenly the vehicle surged upward with tremendous acceleration.

The force pressed us firmly back into our seats as the shuttle climbed rapidly through the atmosphere. Within only a few minutes, the sky outside the windows gradually turned from bright blue to the deep black of space.

Then, almost suddenly, the engines shut down and the intense vibration disappeared. At that moment everything became quiet, and we began floating in weightlessness. That transition—from the powerful acceleration of launch to the peaceful environment of orbit—was one of the most remarkable experiences of my life. It marked the beginning of our journey to conduct scientific research in space and explore the universe beyond our planet. 🚀🌍

Launch of STS-107 from Launch Pad 39A at Kennedy Space Center – Source/Author: NASA - Public Domain

6. What was your first impression when you saw Earth from orbit?

The view of Earth from space is breathtaking. The planet appears incredibly beautiful and fragile, surrounded by the darkness of the universe.

The deployment of the SPARTAN module. Source/Author: NASA - Public Domain

7. What kinds of scientific research did you conduct during your missions?

During my missions aboard the Space Shuttle Columbia, particularly during the STS-87 mission and later the STS-107 mission, our crew conducted a wide range of scientific experiments designed to take advantage of the unique environment of microgravity.

One major area of research involves **materials science**. Scientists were interested in understanding how metals, crystals, and other materials behave when gravity does not influence their formation. These experiments can lead to improvements in manufacturing processes and new types of advanced materials.

Another important field was **fluid physics**. On Earth, gravity strongly affects how liquids move and mix. In microgravity, fluids behave differently, allowing scientists to study fundamental physical processes that are difficult to observe on the ground.

We also conducted **biological and medical experiments**, studying how plants, cells, and microorganisms respond to the conditions of space. These experiments help scientists understand how living systems adapt to weightlessness and may provide insights relevant to medicine and long-duration space travel.

In addition, some experiments focused on **combustion science**, examining how flames behave in microgravity. This research can improve the efficiency and safety of engines and energy systems.

The Space Shuttle served as a unique orbiting laboratory, where scientists from many countries could conduct experiments that expand our understanding of physics, biology, and engineering. For me, participating in these experiments was one of the most rewarding aspects of spaceflight, because the knowledge gained from them continues to benefit both science and life on Earth. 🚀🔬🌍

Mission STS-107 crew in bunk beds on the middeck of the Space Shuttle. Source/Author: NASA – Public Domain

---

8. What message did you hope your journey would send to young people around the world?

One message I hoped my journey would send to young people everywhere is that dreams have no boundaries. No matter where you come from or what challenges you face, curiosity and determination can take you farther than you might imagine.

When I was growing up, I often looked at the night sky and wondered what it would be like to explore space. That sense of curiosity led me to study engineering and eventually to work with NASA. My path was not always easy, but every step brought new opportunities to learn and grow.

I wanted young people, especially students who might think that their dreams are too distant—to see that exploration belongs to everyone. Science, engineering, and discovery are fields open to anyone willing to work hard and pursue knowledge. Space exploration also reminds us that humanity shares a single planet. When you see Earth from orbit, there are no borders visible, only one beautiful world.

If my journey encouraged even a few young people to follow their curiosity, study science, and believe in their possibilities, then it served a purpose beyond the mission itself. The universe is vast, and the next great discoveries will belong to the generation that dares to dream and explore. 🚀🌍✨

---

9. How do you view the future of space exploration?

I believe humanity will continue exploring beyond Earth, returning to the Moon and eventually traveling to Mars. These missions will expand our understanding of the universe.

---

10. What advice would you give to students who dream of becoming astronauts?

My advice to students who dream of becoming astronauts is to follow your curiosity and never stop learning. Space exploration requires strong foundations in science, mathematics, engineering, and technology. These subjects help you understand how the universe works and provide the tools needed to design spacecraft, conduct experiments, and solve complex problems.

At the same time, becoming an astronaut requires more than technical knowledge. It also requires discipline, perseverance, and teamwork. The journey toward space is long, and there will be challenges along the way. What matters most is your determination to continue learning and improving.

Chawla and her husband before the launch of STS-87.

I would also encourage students to explore many interests—aviation, engineering, physics, biology, and even fields like medicine or robotics. Astronaut crews are made up of people with diverse backgrounds who work together to accomplish a mission.

Finally, never let geography or circumstance limit your ambitions. I began my journey as a student who loved airplanes and the night sky, and that curiosity eventually led me to work with NASA and fly aboard the Space Shuttle Columbia.

If you pursue your dreams with passion, dedication, and curiosity, the possibilities are truly limitless. The next generation of explorers may be sitting in classrooms today, looking up at the same stars that once inspired us. 🚀🌍✨

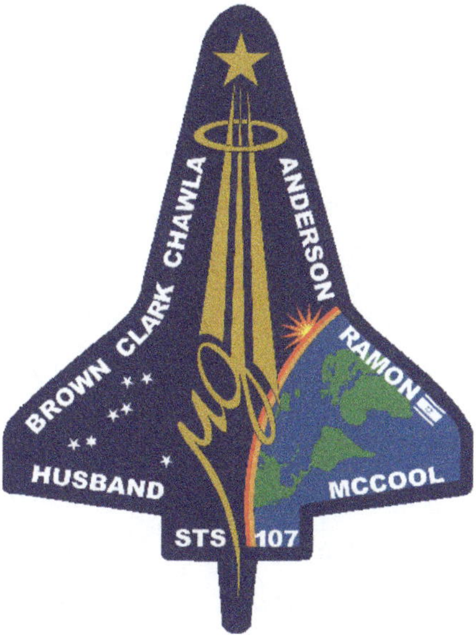

This is the insignia for w:STS-107, which is a multi-discipline microgravity and Earth science research mission with a multitude of international scientific investigations conducted continuously during the planned 16 days on orbit – Public Domain

Interview with Rick Husband

Commander of Space Shuttle Columbia (STS-107)

Introduction

Throughout the history of space exploration, astronauts have represented humanity's courage to venture into the unknown. Among them was Colonel Rick Husband, a man whose dedication to flight, science, and service carried him from the skies of Texas to the frontier of space.

Born in Amarillo, Texas, Rick Husband developed a fascination with aviation at an early age. Watching airplanes soar across the sky sparked a lifelong dream of becoming a pilot. That dream led him to join the United States Air Force, where he became a skilled test pilot and later served as an instructor at the prestigious Air Force Test Pilot School.

His calm leadership, technical expertise, and strong character made him an ideal candidate for NASA's astronaut corps. In 1994, Husband was selected as an astronaut and began training for the Space Shuttle program. His first journey into space came in 1999 aboard the Space Shuttle ***Discovery*** *during mission* ***STS-96****, a mission that helped* *prepare the* ***International Space Station*** *for long-term human habitation.*

Four years later, Husband would command the Space Shuttle ***Columbia*** *on mission* ***STS-107****. Unlike many shuttle flights that focused on construction or satellite deployment, STS-107 was dedicated almost entirely to scientific research. During the sixteen-day mission, astronauts conducted dozens of experiments studying biology, materials science, fluid physics, and the behavior of combustion in microgravity.*

Under Husband's leadership, the crew worked around the clock in two teams, maximizing the scientific output of the mission. The astronauts aboard Columbia formed a diverse international team of explorers and scientists. Among them was mission specialist Kalpana Chawla, whose passion for aerospace engineering inspired millions. The mission itself was remarkably successful. Yet tragedy struck during Columbia's return to Earth.

On February 1, 2003, as the shuttle reentered the atmosphere, damage sustained during launch caused the spacecraft to break apart high above Texas. All seven crew members lost their lives. The loss of Columbia deeply affected NASA and the global space community.

Rick Husband was remembered not only as an exceptional astronaut but also as a devoted husband, father, and man of deep faith. Colleagues often spoke of his humility, kindness, and unwavering dedication to the mission of exploration.

Though his life ended in tragedy, his legacy continues to inspire future astronauts and scientists who pursue humanity's journey into space.

Rick Douglas Husband (July 12, 1957 – February 1, 2003) was an American astronaut and fighter pilot. He traveled into space twice: as pilot of STS-96 and commander of STS-107. Husband and the rest of the crew of STS-107 were killed when *Columbia* disintegrated during reentry into the Earth's atmosphere. He is also a recipient of the Congressional Space Medal of Honor.

---

1. Colonel Husband, when did your interest in aviation begin?

My fascination with flight began when I was a child growing up in Texas. Watching airplanes inspired me to pursue a career in aviation, and eventually I became a pilot in the United States Air Force.

---

2. How did your career in the Air Force prepare you for NASA?

Serving as a pilot and later as a test pilot provided valuable experience operating advanced aircraft and managing complex systems. Those skills are essential for astronauts who must operate spacecraft under challenging conditions.

Pilot (PLT) Rick D. Husband (left) and EVA Specialist Tamara E. Jernigan (right) during STS-96, Source: Wikimedia Commons – Public Domain

3. What did it mean to you to be selected as a NASA astronaut?

Being selected as an astronaut was a tremendous honor. Space exploration represents one of humanity's greatest achievements, and contributing to that effort is a privilege.

---

4. Commander Husband, what do you remember most about your first spaceflight aboard Discovery?

My first spaceflight aboard the Space Shuttle Discovery during the STS-96 mission was an unforgettable experience. As the countdown approached zero, the atmosphere inside the cockpit was calm but intensely focused. Years of training had prepared us for that moment, and every crew member was concentrating on the procedures we had rehearsed countless times.

View of the atmosphere and of the Moon.
Image courtesy of the Image Science & Analysis Laboratory, NASA Johnson Space Center

When the shuttle's main engines ignited, the vehicle began to vibrate with tremendous energy. A few seconds later the solid rocket boosters fired, and suddenly we were accelerating upward with incredible force. You could feel yourself pressed firmly into your seat as the shuttle climbed rapidly through the atmosphere.

Within about eight minutes we reached orbit, and the engines shut down. Almost instantly the intense vibrations stopped, and everything became quiet. Then came the remarkable sensation of weightlessness as objects began floating gently around the cabin.

One of the most memorable moments was looking out the window and seeing Earth from orbit. The view of our planet—its oceans, clouds, and continents—was breathtaking. The mission itself was also historic because we were delivering supplies and equipment to the International Space Station, helping prepare the station for future crews.

For me, that first flight was both humbling and inspiring. It reminded me how extraordinary it is that human beings can work together to explore space and build structures orbiting hundreds of kilometers above our planet. 🚀🌍

---

5. As commander of Columbia, what was your primary responsibility during the STS-107 mission?

As commander of the Space Shuttle Columbia during the STS-107 mission, my primary responsibility was the overall safety and success of the mission and the well-being of the entire crew. The commander oversees every phase of the flight—from launch and orbital operations to reentry and landing. That includes working closely with the pilot and mission specialists to ensure that all systems of the spacecraft operate correctly and that the mission objectives are completed safely.

STS-107 was primarily a **scientific research mission**, and our crew conducted a large number of experiments inside the Spacehab laboratory module. My role was to coordinate the activities of the crew, maintain communication with mission control, and make sure that the shuttle's systems were functioning properly throughout the mission.

A view of Mount Fuji and the surrounding area from Columbia

Another important responsibility was **team leadership**. A shuttle crew is made up of people from different scientific and technical backgrounds, and the commander helps ensure that everyone works together smoothly while maintaining the demanding schedule of experiments and operations.

Ultimately, the commander must always be prepared to make critical decisions if unexpected situations arise. Spaceflight requires constant awareness, careful planning, and a deep trust between the crew and the teams on the ground at NASA.

Leading the mission was both an honor and a responsibility, representing the efforts of thousands of engineers, scientists, and technicians who worked together to make human space exploration possible.

---

6. What made the STS-107 mission unique?

The STS-107 mission aboard the Space Shuttle Columbia was unique because it was dedicated almost entirely to scientific research in microgravity. Unlike many shuttle missions that focused on building or servicing the International Space Station, our mission functioned as a fully operational laboratory in orbit. Inside the Spacehab research module, we conducted **more than 80 scientific experiments** from researchers around the world.

These experiments covered a wide range of fields, including **biology, medicine, materials science, fluid physics, and combustion science**. Scientists were particularly interested in studying how physical and biological processes behave in microgravity, where gravity does not interfere with the results.

Another unique aspect of the mission was the continuous research schedule. The crew was divided into two teams that worked in shifts, allowing experiments to operate nearly twenty-four hours a day during the mission.

STS-107 also represented a strong example of international scientific cooperation, with research contributions from many countries.

In many ways, the mission demonstrated the remarkable potential of space as a laboratory. By conducting experiments in orbit, scientists can explore fundamental questions that help improve technology, medicine, and our understanding of nature itself. 🚀🔬🌍

Spacehab's Research Double Module in Columbia's payload bay during STS-107 – Public Domain

7. How did working with an international team influence the mission?

Working with an international crew during the STS-107 mission aboard the Space Shuttle Columbia was one of the most rewarding aspects of the mission. Our crew included astronauts and payload specialists from different countries and cultural backgrounds. Each person brought unique expertise, perspectives, and experiences to the mission. This diversity strengthened the team and contributed to the success of the scientific work we conducted.

One of the most important examples was the participation of Kalpana Chawla and Ilan Ramon, whose presence reflected the global nature of modern space exploration. Scientists from many countries contributed experiments to the Spacehab research module, so in many ways the mission represented the efforts of an international scientific community.

Working together in space reinforces an important truth: when you look at Earth from orbit, you cannot see national borders. You see one planet shared by all humanity. Space exploration often brings people together in ways that transcend politics or geography. Cooperation between nations allows us to combine knowledge, technology, and creativity to accomplish goals that no single country could achieve alone.

In that sense, missions like STS-107 demonstrate that exploration of space is not only a technological achievement, but also a powerful example of international collaboration in the pursuit of knowledge.

---

8. What does seeing Earth from space teach us about our planet?

From orbit, Earth appears incredibly beautiful and fragile. It reminds you that our planet is a shared home that we must protect.

Husband's Camaro on display at Rick Husband Amarillo International Airport
Source: Wikimedia Commons- CC BY-SA 4.0

---

9. What advice would you give to young people who dream of becoming astronauts?

Work hard, pursue your education, and never give up on your dreams. Exploration requires dedication and perseverance.

10. What message would you share with future explorers who will travel beyond Earth?

Exploration is part of the human spirit. Each generation builds upon the achievements of those who came before. The future of space exploration will belong to those who continue reaching for the stars.

Statue of Rick Husband at Amarillo International Airport
Source: Wikimedia Commons – CC BY-SA 4.0

Epilogue

Humanity's Next Giant Leap

The story of human space exploration is still in its earliest chapters. In the span of just a few decades, humanity progressed from the first fragile rockets to walking on the Moon, building laboratories in orbit, and sending robotic explorers to the farthest reaches of our solar system. These achievements were once considered impossible. Yet they were realized through the courage of astronauts and cosmonauts who dared to leave Earth behind.

The pioneers whose voices appear in this book belonged to a remarkable generation. They flew into space when technology was uncertain and when every mission carried the possibility of failure. Their journeys were acts of both scientific progress and profound personal bravery. Some returned safely, their accomplishments celebrated across the world. Others lost their lives while pushing the boundaries of human exploration. Their sacrifices remind us that progress often requires extraordinary courage.

Yet their legacy did not end with the missions they flew. Today, a new era of space exploration is emerging. Governments and private companies are developing powerful new rockets, preparing to return humans to the Moon and eventually send astronauts to Mars. International cooperation has become a defining feature of modern spaceflight, as astronauts from many nations live and work together aboard the International Space Station.

Future explorers may establish permanent bases on the Moon, travel to Mars, and venture farther into the solar system than any human has gone before. One day, humanity may even send missions beyond our solar neighborhood to explore distant stars. These ambitions are possible only because of the pioneers who came before.

The astronauts and cosmonauts remembered in this book demonstrated that humans can survive and thrive in space. They proved that exploration does not end at the edge of our atmosphere. Their courage expanded the boundaries of what humanity believed possible. But perhaps the greatest lesson of space exploration is the perspective it offers.

Astronauts who looked down upon Earth from orbit often described a profound realization: our planet is small, fragile, and deeply interconnected. From space, national borders disappear, conflicts seem trivial, and the shared destiny of humanity becomes clearer. Seen from space, Earth is not divided into nations, it is a single world.

This perspective may be one of the most valuable gifts astronauts have given humanity. The pioneers of space exploration opened the door to a future that stretches far beyond our planet. Their journeys remind us that the desire to explore is one of the defining traits of our species. The stars have always inspired humanity to dream. Now, for the first time in history, we have begun the journey toward them.

Author's Reflection

Artificial Intelligence, Memory, and the Voices of Exploration

Human space exploration is one of the most remarkable achievements in the history of civilization. In just a few decades, humanity moved from the first fragile rockets to walking on the Moon, building laboratories in orbit, and sending spacecraft deep into the solar system. These extraordinary accomplishments were made possible by the courage of astronauts and cosmonauts who accepted the risks of exploration to expand human knowledge.

The individuals featured in this book represent some of the most inspiring pioneers of the Space Age. Their journeys reflect the best qualities of humanity—curiosity, determination, ingenuity, and the willingness to explore the unknown. Yet as time passes, many of these explorers are no longer with us.

Artificial intelligence offers a new and fascinating way to remember their voices and reflect upon their experiences. By combining historical documents, interviews, memoirs, and scientific records, AI can help reconstruct thoughtful conversations that capture the spirit and ideas of these remarkable individuals.

The interviews presented in this book are therefore not literal transcripts. Instead, they are AI-assisted imaginative reconstructions, built upon historical facts and the documented words and actions of the astronauts themselves. They are designed to provide readers with a sense of how these pioneers might reflect upon their missions, their motivations, and the meaning of exploration.

In this sense, artificial intelligence becomes a bridge between history and imagination. Just as telescopes allow us to see distant galaxies and spacecraft allow us to travel beyond Earth, AI allows us to explore historical perspectives in new ways. It enables us to ask questions across time—questions that help illuminate the experiences of those who ventured into space during one of humanity's most transformative eras.

The astronauts featured in this book came from different nations, cultures, and backgrounds. Some flew during the intense rivalry of the Cold War, when the United States and the Soviet Union competed to achieve the first milestones in space. Others participated in the era of international cooperation that followed, when astronauts from many nations worked together aboard spacecraft and space stations.

Despite these differences, they all shared a common belief: that exploration is an essential part of human progress. Their missions expanded scientific knowledge, inspired generations of students, and demonstrated what humanity can achieve when science, engineering, and imagination work together.

Artificial intelligence represents another frontier of discovery. Just as the pioneers of space exploration opened a path beyond Earth, advances in AI are opening new possibilities for understanding knowledge, creativity, and human history. Both frontier space and artificial intelligence—challenge us to think beyond traditional boundaries and imagine what the future might hold.

As humanity prepares for new missions to the Moon, Mars, and perhaps even more distant destinations, the stories of the astronauts in this book remind us that exploration is never the work of a single individual. It is a collective achievement that involves scientists, engineers, teachers, and dreamers working together across generations. The pioneers of the Space Age showed us that the horizon of human possibility extends far beyond our planet. Their courage continues to inspire us. And their voices imagined here through the lens of artificial intelligence—remind us that the journey to the stars has only just begun. 🚀

About the Interviewer

Dr. Borko Furht is a leading scholar, innovator, and author whose work spans the frontiers of artificial intelligence, multimedia systems, and next-generation communication technologies. A Professor of Computer Science and Engineering at Florida Atlantic University, he serves as Director and Principal Investigator of the NSF Industry/University Cooperative Research Center for Advanced Knowledge Enablement (CAKE)—a national research consortium that has partnered with more than fifty industry members and produced groundbreaking advances across AI, data science, video coding, and machine-to-machine communication.

Over the course of his distinguished career, Dr. Furht has authored and edited over 40 books, published more than 300 scientific papers, and holds more than 100 U.S. and international patents—including 13 essential patents for the

H.266/VVC video coding standard. His research initiatives have been supported by the National Science Foundation, National Institutes of Health, NASA, Office of Naval Research, and major industry partners such as Google, Motorola, IBM, GA Telesis, LexisNexis, and OP Solutions. He is also the founding Editor-in-Chief of Springer's Multimedia Tools and Applications Journal (est. 1994) and co-founder of the Journal of Big Data (est. 2011).

Beyond his academic work, Dr. Furht is a passionate creator who explores how technology can illuminate human stories. His interest in the intersection of AI and cultural history led to the development of *AI Interviews with Astronauts Legends*, a book that blends imagination, biography, and machine intelligence to give new voice to extraordinary figures of the past. His creative projects aim to make technology accessible, meaningful, and emotionally resonant for readers of all ages.

Dr. Furht is recognized internationally for his contributions to research and education, receiving numerous awards including FAU's Presidential Award for Career Achievements (PACA, 2024), Researcher of the Year Awards (2013, 2019, and 2026) and election to the prestigious Academia Europaea in 2019. His work continues to inspire students, researchers, and readers around the world.

He lives in Boca Raton, Florida, where he divides his time between research, writing, mentorship, and exploring new ways that AI can enrich human understanding and creativity.

My Personal Connection to the Space Shuttle Program

Designing a New Launch Computer Control Architecture for NASA

My interest in space exploration has not been limited to studying the achievements of astronauts and engineers. Earlier in my career, I had the privilege of contributing directly to the technology supporting NASA's Space Shuttle program.

Around 1990, while I was serving as **Vice President of Research at MODCOMP in Fort Lauderdale, Florida**, I worked with **Dr. Robert Luken of NASA's Kennedy Space Center** on the design and evaluation of a new computer architecture for the **Data Acquisition and Space Shuttle Launch Control System**. Our work resulted in the technical paper *"The Space Shuttle Launch Computer Control System at NASA Kennedy Space Center,"* authored by Robert Luken and me. The published paper identifies my affiliation as MODCOMP and Dr. Luken's as NASA Kennedy Space Center.

The challenge was formidable. At the time, the Shuttle launch-control environment relied on approximately **400 MODCOMP 16-bit proprietary computers**, and NASA was investigating how this system could eventually be modernized using state-of-the-art, open, real-time multiprocessor technology. The launch-control system performed critical functions beginning approximately **72 hours before launch and continuing until T-minus 31 seconds**, when primary control was transferred to the Space

Shuttle's onboard computers. Even after that transfer, the ground system continued monitoring the Shuttle.

Our proposed architecture was based on a **symmetric, tightly coupled multiprocessor system** with shared high-speed memory, dedicated high-performance I/O processors, efficient interrupt handling, and a multiprocessor real-time UNIX operating system. The objective was to combine substantial computing power with the deterministic response and reliability required for one of the world's most demanding real-time applications.

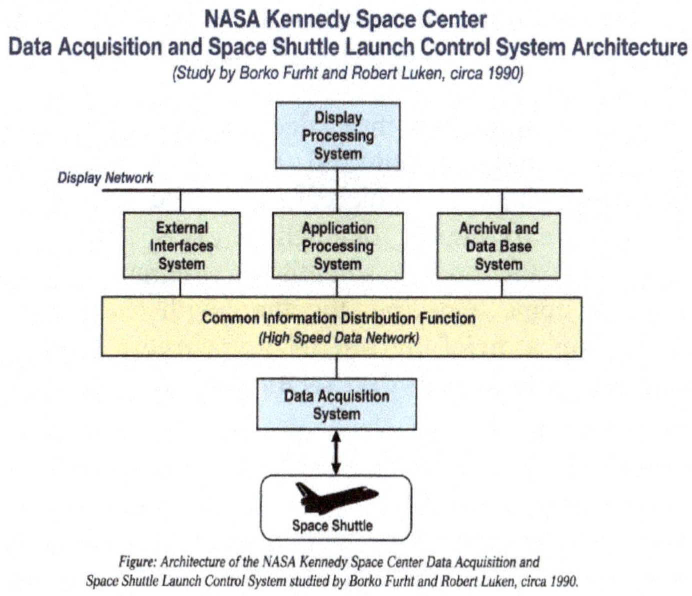

Architecture of the NASA Kennedy Space Center Data Acquisition and Space Shuttle Launch Control System studied by Borko Furht and Robert Luken, circa 1990.

The scale of the problem was remarkable for that era. The Data Acquisition System was expected to process approximately **60,000 significant data changes every**

second, representing about **577,000 bytes per second** of front-end traffic. Our analysis showed that the Data Acquisition Processor required approximately **23 million instructions per second**, and we mapped its functions onto five processing elements—**three dedicated I/O processors and two CPUs**.

A prototype of the tightly coupled multiprocessor architecture was subsequently developed at MODCOMP using multiple Motorola MC68030 processors. The paper also described the planned **REAL/STAR System**, designed to support from 2 to 20 MC88100 RISC processors and MODCOMP's multiprocessor real-time **REAL/IX** operating system.

Looking back more than three decades later, this project remains particularly meaningful to me. The astronauts described throughout this book represented the visible human face of space exploration, but behind every launch stood thousands of engineers, scientists, programmers, and computer systems working together with extraordinary precision. **For a brief period of my career, I had the privilege of being part of that technological world.**

Acknowledgments

The creation of this book has been both an intellectual journey and a deeply personal one. Although the voices brought to life in these pages belong to astronauts from different nations, missions, and eras of space exploration, the process of imagining their reflections has been profoundly collaborative.

I would first like to express my gratitude to my colleagues and students at Florida Atlantic University, whose curiosity, insight, and enthusiasm continually inspire my work. The vibrant research culture of the **NSF Center for Advanced Knowledge Enablement (CAKE)** has shaped my thinking about artificial intelligence, creativity, and the future of human–machine interaction. I am especially grateful to the dedicated researchers who have joined me in exploring how AI can help us revisit history and illuminate the stories of extraordinary individuals who expanded humanity's horizons.

I extend sincere appreciation to the **National Science Foundation**, whose long-standing support through multiple research initiatives has enabled the exploration of technologies that form the intellectual foundation for projects like this one. Their commitment to discovery and innovation continues to advance our understanding of artificial intelligence and its role in preserving and interpreting human achievements.

To the remarkable astronauts whose lives and missions are honored in this book—pioneers who left Earth to explore the vastness of space—I offer my deepest respect. These men and women carried humanity's dreams beyond the atmosphere. They faced unimaginable risks, endured isolation and uncertainty, and helped transform the impossible into reality. From the first steps into orbit to the exploration of the Moon and beyond, their courage, discipline, and determination opened a new chapter in human history. It has been both a privilege and an honor to explore their legacies through the lens of modern technology, bringing their stories of exploration, sacrifice, and discovery into conversation with our own time.

I am profoundly grateful to my wife **Sandra**, my daughter **Tanya**, and my granddaughter **Sophia**, whose encouragement, patience, and unwavering support have accompanied me through every chapter of my career. I also thank my wife **Sandra** and my brother **Sasha**, whose suggestion to explore interviews with legendary astronauts helped inspire the creation of this book.

I must also acknowledge my quiet, tireless collaborator — the AI Assistant. Though it has no heartbeat, it listens with patience and responds with unexpected depth. Through its algorithms, I found not only the echoes of historical voices, but reflections of humanity's enduring curiosity about the universe.

Finally, I thank the readers—those who are fascinated by space exploration, those who embrace the possibilities of artificial intelligence, and those who believe that the stories of the men and women who ventured beyond Earth deserve to be remembered and retold. This book is dedicated to the idea that exploration never truly ends; it continues wherever curiosity meets courage.

With gratitude,
Borko Furht
Boca Raton, Florida

Image and Copyright Note

Images reproduced in this book were obtained primarily from Wikimedia Commons and other publicly accessible archival sources. Copyright and licensing information was reviewed on the individual Wikimedia Commons file-description pages. Images identified as public domain or CC0 are reproduced on that basis. Certain images are made available under Creative Commons licenses, including CC BY and CC BY-SA, and are reproduced subject to the terms of their respective licenses. Image credits, creators, sources, and applicable licenses are acknowledged where required.

Appendix A

Timeline of Major Space Exploration Milestones (1957–Present)

The following timeline highlights some of the most significant milestones in the history of human space exploration. From the launch of the first artificial satellite to modern missions preparing for journeys to the Moon and Mars, these events illustrate the remarkable progress humanity has made in exploring space.

1957 The Soviet Union launches **Sputnik 1**, the first artificial satellite to orbit Earth, marking the beginning of the Space Age.

1961 Yuri Gagarin becomes the first human to travel into space and orbit Earth aboard **Vostok 1**.

1961 Alan Shepard becomes the first American in space aboard **Freedom 7**.

1962 John Glenn becomes the first American to orbit Earth during the **Friendship 7** mission.

1963 Valentina Tereshkova becomes the first woman in space aboard **Vostok 6**.

1965 Alexei Leonov performs the first spacewalk (extravehicular activity) during the **Voskhod 2** mission.

1968 Apollo 8 becomes the first crewed spacecraft to orbit the Moon, providing humanity with the famous "Earthrise" view.

1969 Neil Armstrong and Buzz Aldrin become the first humans to walk on the Moon during the **Apollo 11** mission.

1971 The Soviet Union launches **Salyut 1**, the world's first space station.

1975 The **Apollo–Soyuz Test Project** marks the first international crewed space mission between the United States and the Soviet Union.

1981 NASA launches the first **Space Shuttle mission (Columbia)**, beginning a new era of reusable spacecraft.

1986 The Soviet Union launches **Mir**, the first modular space station that supports long-duration human habitation.

1998 Construction of the **International Space Station (ISS)** begins through cooperation among multiple nations.

2000 The ISS becomes permanently inhabited by astronauts from around the world.

2004 SpaceShipOne becomes the first privately funded spacecraft to carry a human into suborbital space.

2012 The **Curiosity rover** successfully lands on Mars, dramatically advancing robotic exploration of the planet.

2020 NASA and SpaceX launch **Crew Dragon**, returning human orbital launches to the United States after nearly a decade.

2021 NASA's **Perseverance rover** lands on Mars and begins searching for signs of ancient life.

2022 NASA launches the **Artemis I mission**, a major step toward returning astronauts to the Moon.

2024–Present The **Artemis program** prepares for a new era of human lunar exploration, including plans for a permanent presence on the Moon and future missions to Mars.

Looking Ahead

Humanity now stands at the threshold of a new age of exploration. Future missions aim to establish lunar bases, explore Mars with human crews, and develop technologies that may eventually allow journeys to distant worlds beyond our solar system.

The milestones in this timeline represent only the beginning of humanity's journey into space. The greatest discoveries—and the next giant leaps—still lie ahead. 🚀

Appendix B

Key Space Missions That Changed History

Throughout the history of space exploration, certain missions have fundamentally changed humanity's understanding of the universe and our place within it. These missions introduced new technologies, expanded scientific knowledge, and demonstrated the courage and ingenuity required to explore beyond Earth.

The following missions represent some of the most influential milestones in the development of human space exploration.

Sputnik 1 (1957)

Launched by the Soviet Union, **Sputnik 1** became the first artificial satellite to orbit Earth. Its successful launch marked the beginning of the Space Age and initiated the global competition known as Space Race. The mission demonstrated that artificial satellites could remain in orbit and transmit signals to Earth.

Vostok 1 (1961)

During this historic mission, Yuri Gagarin became the first human to travel into space and orbit Earth. His 108-minute journey aboard Vostok 1 proved that humans could survive in space and opened the door to future crewed missions.

Freedom 7 (1961)

Alan Shepard became the first American astronaut in space during the Mercury-Redstone 3 mission. Although the flight was suborbital and lasted only about fifteen minutes, it represented a crucial step in the United States' early human spaceflight program.

Friendship 7 (1962)

John Glenn completed three orbits around Earth aboard the Friendship 7 spacecraft. This mission demonstrated the United States' capability for orbital human spaceflight and strengthened public confidence in NASA's Mercury program.

Voskhod 2 (1965)

During the Voskhod 2 mission, Alexei Leonov performed the first spacewalk in history. Floating outside the spacecraft for approximately twelve minutes, Leonov demonstrated that astronauts could work outside their vehicles in the vacuum of space.

Apollo 8 (1968)

Apollo 8 became the first crewed spacecraft to orbit the Moon. Astronauts aboard the mission witnessed the famous "Earthrise" view and demonstrated that humans could travel safely to the Moon and return to Earth.

Apollo 11 (1969)

Apollo 11 achieved one of the greatest accomplishments in human history when Neil Armstrong and Buzz Aldrin became the first humans to walk on the surface of the Moon. Armstrong's famous words, "one small step for man, one giant leap for mankind," captured the significance of the moment.

Apollo 13 (1970)

Although the mission never landed on the Moon, Apollo 13 became one of NASA's most remarkable demonstrations of teamwork and problem-solving. After an oxygen tank explosion crippled the spacecraft, the crew—including Jim Lovell—successfully returned to Earth thanks to innovative solutions developed by both astronauts and engineers on the ground.

Salyut 1 (1971)

The Soviet Union launched **Salyut 1**, the world's first space station. This mission demonstrated that humans could live and conduct scientific research in space for extended periods.

Apollo–Soyuz Test Project (1975)

This joint mission between the United States and the Soviet Union marked the first international human spaceflight collaboration. The successful docking of the Apollo and Soyuz spacecraft symbolized a moment of cooperation during the Cold War.

Space Shuttle Program (1981–2011)

NASA's Space Shuttle program introduced reusable spacecraft capable of launching satellites, conducting scientific experiments, and building large structures in orbit. The program played a critical role in constructing the International Space Station and advancing human spaceflight.

Mir Space Station (1986–2001)

The Soviet and later Russian **Mir** space station became the first long-term research laboratory in orbit. Astronauts and cosmonauts lived aboard Mir for extended missions that greatly expanded knowledge of long-duration spaceflight.

International Space Station (1998–Present)

The **International Space Station (ISS)** represents one of the largest international scientific collaborations in history. Astronauts from many nations have lived and worked aboard the ISS, conducting experiments in microgravity and advancing our understanding of space medicine, physics, and Earth science.

Mars Curiosity Rover (2012)

NASA's **Curiosity rover** landed on Mars and began a comprehensive scientific investigation of the planet's geology and climate. The mission confirmed that Mars once possessed environmental conditions capable of supporting microbial life.

Artemis Program (2020s–Present)

NASA’s **Artemis program** aims to return astronauts to the Moon and establish a sustainable human presence there. These missions are designed to prepare humanity for future exploration of Mars and deeper space.

These missions illustrate the extraordinary progress humanity has achieved in exploring space. Each mission built upon the successes and lessons of those that came before, gradually transforming spaceflight from a daring experiment into a sustained global endeavor.

The missions of the future will continue this legacy, carrying humanity farther into the cosmos than ever before. 🚀

Appendix C

Glossary of Spaceflight and Astronautics Terms

This glossary provides brief explanations of important terms commonly used in space exploration, astronautics, and human spaceflight.

Apollo Program
NASA's human spaceflight program that successfully landed astronauts on the Moon between 1969 and 1972.

Astronaut
A person trained to travel and perform missions in space. In the United States and many Western countries, space travelers are called astronauts, while in Russia they are known as cosmonauts.

Booster Rocket
A powerful rocket stage that provides the initial thrust needed to launch a spacecraft into space.

Capsule

A small, pressurized spacecraft designed to transport astronauts and safely return them to Earth.

Command Module

The primary spacecraft component used by astronauts during Apollo missions, containing navigation controls, living space, and reentry systems.

Cosmonaut

A Russian or Soviet astronaut trained for human spaceflight missions.

Docking

The process of connecting two spacecraft together in orbit so that astronauts and materials can transfer between them.

Extravehicular Activity (EVA)

Any activity performed by astronauts outside their spacecraft while in space. EVA is commonly known as a spacewalk.

G-Force (Gravitational Force)

The force experienced by astronauts during acceleration, such as during rocket launch or reentry.

International Space Station (ISS)

A large space station orbiting Earth where astronauts from many countries live and conduct scientific experiments.

Launch Window

A specific period during which a spacecraft must be launched to reach its intended destination.

Low Earth Orbit (LEO)

An orbit around Earth at an altitude typically between 160 and 2,000 kilometers above the planet's surface.

Lunar Module

The spacecraft used during Apollo missions to transport astronauts from lunar orbit to the surface of the Moon.

Microgravity

The condition in which objects appear weightless because they are in continuous free fall while orbiting Earth.

Mission Control
The ground-based team of engineers and specialists responsible for monitoring and guiding spacecraft during a mission.

Orbital Mechanics
The scientific study of how objects move in space under the influence of gravity.

Payload
The cargo carried by a spacecraft, which may include satellites, scientific instruments, or astronauts.

Reentry
The process by which a spacecraft returns from space and passes through Earth's atmosphere.

Rocket
A vehicle powered by engines that expel high-speed gases to produce thrust, allowing it to travel into space.

Space Shuttle
NASA's reusable spacecraft used between 1981 and 2011 to transport astronauts and cargo to orbit.

Spacewalk
An activity in which astronauts leave their spacecraft and work outside in space while wearing specialized suits.

Spacecraft
Any vehicle designed to travel or operate in outer space.

Telemetry
The process of transmitting data from a spacecraft to Earth for monitoring and analysis.

Trajectory
The path a spacecraft follows through space.

Zero Gravity
A state in which objects experience weightlessness due to continuous free fall in orbit.

These terms form the vocabulary of modern space exploration and help explain the technologies and experiences that enable humans to travel beyond Earth.

www.ingramcontent.com/pod-product-compliance
Lightning Source LLC
Chambersburg PA
CBHW071222230926
51849CB00001B/7
9786195287525